Methods in Molecular Biology

Series Editor
John M. Walker
School of Life and Medical Sciences
University of Hertfordshire
Hatfield, Hertfordshire, UK

For further volumes:
http://www.springer.com/series/7651

For over 35 years, biological scientists have come to rely on the research protocols and methodologies in the critically acclaimed *Methods in Molecular Biology* series. The series was the first to introduce the step-by-step protocols approach that has become the standard in all biomedical protocol publishing. Each protocol is provided in readily-reproducible step-by-step fashion, opening with an introductory overview, a list of the materials and reagents needed to complete the experiment, and followed by a detailed procedure that is supported with a helpful notes section offering tips and tricks of the trade as well as troubleshooting advice. These hallmark features were introduced by series editor Dr. John Walker and constitute the key ingredient in each and every volume of the *Methods in Molecular Biology* series. Tested and trusted, comprehensive and reliable, all protocols from the series are indexed in Pub Med.

Stem Cell Niche

Methods and Protocols

Second Edition

Edited by

Kursad Turksen

Ottawa Hospital Research Institute, Ottawa, ON, Canada

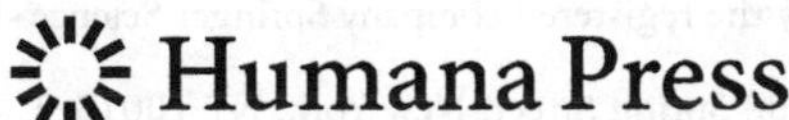

Editor
Kursad Turksen
Ottawa Hospital Research Institute
Ottawa, ON, Canada

ISSN 1064-3745 ISSN 1940-6029 (electronic)
Methods in Molecular Biology
ISBN 978-1-4939-9507-3 ISBN 978-1-4939-9508-0 (eBook)
https://doi.org/10.1007/978-1-4939-9508-0

This Humana imprint is published by the registered company Springer Science+Business Media, LLC, part of Springer Nature.
The registered company address is: 233 Spring Street, New York, NY 10013, U.S.A.

Preface

The idea that there is a specialized microenvironment—the so-called stem cell niche—that regulates hematopoietic stem cell function was first proposed by Ray Schofield over three decades ago. As interest in stem cell biology has exploded over the last 10 years, so too has interest in the stem cell niche. Over the last several years, intense effort has been underway to characterize the cellular-molecular-biochemical elements of not only the hemopoietic stem cell niche but also the niches for other stem cells. To recognize and build upon the advances achieved since the initial volume on the topic of stem cell niches, I have collected a series of updates as well as new protocols for this new volume.

The protocols gathered here are faithful to the mission statement of the *Methods in Molecular Biology* series. They are well established and described in an easy-to-follow, step-by-step fashion so as to be valuable for not only experts but also novices in the stem cell field. That goal is achieved because of the generosity of the contributors who have carefully described their protocols in this volume, and I am grateful for their efforts.

My thanks as well go to Dr. John Walker, the Editor in Chief of the *Methods in Molecular Biology* series, for giving me the opportunity to create this volume and for supporting me along the way.

I am also grateful to Patrick Marton, the Executive Editor of *Methods in Molecular Biology* and Springer Protocols series, for his continuous support from idea to completion of this volume.

I would like to thank David C. Casey, Senior Editor for *Methods in Molecular Biology*, for his outstanding editorial work during the production of this volume.

Finally, I would like to thank the production crew for putting together an outstanding volume.

Ottawa, ON, Canada ***Kursad Turksen***

Contents

Contributors

Cristina C. Barrias • *i3S – Instituto de Investigação e Inovação em Saúde, University of Porto, Porto, Portugal; INEB – Instituto de Engenharia Biomédica, University of Porto, Porto, Portugal; ICBAS – Instituto de Ciências Biomédicas Abel Salazar, University of Porto, Porto, Portugal*

Chandra S. Bhol • *Department of Life Science, National Institute of Technology Rourkela, Rourkela, Odisha, India*

Sujit K. Bhutia • *Department of Life Science, National Institute of Technology Rourkela, Rourkela, Odisha, India*

Sílvia J. Bidarra • *i3S – Instituto de Investigação e Inovação em Saúde, University of Porto, Porto, Portugal; INEB – Instituto de Engenharia Biomédica, University of Porto, Porto, Portugal*

Nathalie Brouard • *Unité Mixte de Recherche - S1255 Recherche Plaquettes Sanguines: Hémostase, Thrombose, Transfusion, Université de Strasbourg, INSERM, Etablissement Français du Sang-Alsace, Fédération de Médecine Translationnelle de Strasbourg, Strasbourg, France*

Madeline Campbell • *Sydney Medical School, University of Sydney, Sydney, NSW, Australia*

Susana Cardoso • *Instituto de Engenharia de Sistemas de Computadores – Microsystems and Nanotechnology (INESC–MN), Lisboa, Portugal*

Mamta Chabria • *Roche Pharma Research and Early Development, Roche Innovation Center Basel, F. Hoffmann-La Roche Ltd., Basel, Switzerland*

Bénédicte Chazaud • *Institut NeuroMyoGène, Univ Lyon, Université Claude Bernard Lyon 1, INSERM U1217, CNRS UMR5310, Lyon, France*

Marta H. G. Costa • *Department of Bioengineering and IBB-Institute for Bioengineering and Biosciences, Instituto Superior Técnico, Universidade de Lisboa, Lisboa, Portugal*

Cláudia L. da Silva • *Department of Bioengineering and IBB-Institute for Bioengineering and Biosciences, Instituto Superior Técnico, Universidade de Lisboa, Lisboa, Portugal; The Discoveries Centre for Regenerative and Precision Medicine, Lisbon Campus, Instituto Superior Técnico, Universidade de Lisboa, Lisboa, Portugal*

Beatriz de Lucas • *Universidad Europea de Madrid, Madrid, Spain; Instituto de Investigación Hospital 12 de Octubre (I+12), Madrid, Spain*

Matthew R. Eber • *Department of Cancer Biology and Comprehensive Cancer Center, Wake Forest School of Medicine, Winston-Salem, NC, USA*

David M. Feliciano • *Department of Biological Sciences, Clemson University, Clemson, SC, USA*

Frederico Castelo Ferreira • *Department of Bioengineering and IBB-Institute for Bioengineering and Biosciences, Instituto Superior Técnico, Universidade de Lisboa, Lisboa, Portugal; The Discoveries Centre for Regenerative and Precision Medicine, Lisbon Campus, Instituto Superior Técnico, Universidade de Lisboa, Lisboa, Portugal*

Gemma A. Figtree • *Sydney Medical School, University of Sydney, Sydney, NSW, Australia*

Beatriz G. Gálvez • *Universidad Europea de Madrid, Madrid, Spain; Instituto de Investigación Hospital 12 de Octubre (I+12), Madrid, Spain*

CARMINE GENTILE • *Sydney Medical School, University of Sydney, Sydney, NSW, Australia; Beth Israel Deaconess Medical Center, Harvard Medical School, Boston, MA, USA; Kolling Institute, Royal North Shore Hospital, Sydney, NSW, Australia*

JOYDEEP GHOSH • *Department of Medicine, Indiana University School of Medicine, Indianapolis, IN, USA*

AVIVA J. GOEL • *Department of Cell, Developmental, and Regenerative Biology, Icahn School of Medicine at Mount Sinai, New York, NY, USA*

CHENG-HU HU • *State Key Laboratory of Military Stomatology, Center for Tissue Engineering, School of Stomatology, Fourth Military Medical University, Xi'an, Shaanxi, China; Research and Development Center for Tissue Engineering, Fourth Military Medical University, Xi'an, Shaanxi, China; Xi'an Institute of Tissue Engineering and Regenerative Medicine, Xi'an, Shaanxi, China*

DAVID JOHANNES HUELS • *LEXOR, Center for Experimental and Molecular Medicine, Amsterdam UMC, University of Amsterdam, Cancer Center Amsterdam, Amsterdam, Netherlands; Oncode Institute, Amsterdam, Netherlands*

YAN JIN • *State Key Laboratory of Military Stomatology, Center for Tissue Engineering, School of Stomatology, Fourth Military Medical University, Xi'an, Shaanxi, China; Research and Development Center for Tissue Engineering, Fourth Military Medical University, Xi'an, Shaanxi, China; Xi'an Institute of Tissue Engineering and Regenerative Medicine, Xi'an, Shaanxi, China*

JOAQUIM M. S. CABRAL • *Department of Bioengineering and IBB-Institute for Bioengineering and Biosciences, Instituto Superior Técnico, Universidade de Lisboa, Lisboa, Portugal; The Discoveries Centre for Regenerative and Precision Medicine, Lisbon Campus, Instituto Superior Técnico, Universidade de Lisboa, Lisboa, Portugal*

LAURA KEROSUO • *Department of Health and Human Services, Neural Crest Development and Disease Unit, National Institute of Dental and Craniofacial Research, National Institutes of Health, Bethesda, MD, USA; Biochemistry and Developmental Biology, Faculty of Medicine, University of Helsinki, Helsinki, Finland; Single-Cell Quantitative Biology Lab, Department of Biochemistry and Developmental Biology, Faculty of Medicine, University of Helsinki, Helsinki, Finland; Neural Crest Development and Disease Unit, National Institute for Dental and Craniofacial Research, National Institutes of Health, Bethesda, MD, USA*

ALEKSANDAR BURYANOV KIROV • *LEXOR, Center for Experimental and Molecular Medicine, Amsterdam UMC, University of Amsterdam, Cancer Center Amsterdam, Amsterdam, Netherlands; Oncode Institute, Amsterdam, Netherlands*

ROBERT S. KRAUSS • *Department of Cell, Developmental, and Regenerative Biology, Icahn School of Medicine at Mount Sinai, New York, NY, USA*

CLAIRE LATROCHE • *San Raffaele Telethon Institute for Gene Therapy, Milan, Italy*

ANTTI LIGNELL • *Single-Cell Quantitative Biology Lab, Department of Biochemistry and Developmental Biology, Faculty of Medicine, University of Helsinki, Helsinki, Finland*

KEWAL K. MAHAPATRA • *Department of Life Science, National Institute of Technology Rourkela, Rourkela, Odisha, India*

JAN PAUL MEDEMA • *LEXOR, Center for Experimental and Molecular Medicine, Amsterdam UMC, University of Amsterdam, Cancer Center Amsterdam, Amsterdam, Netherlands; Oncode Institute, Amsterdam, Netherlands*

SAFA F. MOHAMAD • *Department of Microbiology and Immunology, Indiana University School of Medicine, Indianapolis, IN, USA*

SOFIE MOHLIN • *Division of Pediatrics, Department of Clinical Sciences, Lund University, Lund, Sweden*
TIAGO S. MONTEIRO • *Instituto de Engenharia de Sistemas de Computadores – Microsystems and Nanotechnology (INESC–MN), Lisboa, Portugal*
MARY C. MORTON • *Department of Biological Sciences, Clemson University, Clemson, SC, USA*
PRAJNA P. NAIK • *Department of Life Science, National Institute of Technology Rourkela, Rourkela, Odisha, India; PG Department of Zoology, Vikram Deb (Auto) College, Jeypore, Odisha, India*
VICTORIA N. NECKLES • *Department of Biological Sciences, Clemson University, Clemson, SC, USA*
DEBASNA P. PANIGRAHI • *Department of Life Science, National Institute of Technology Rourkela, Rourkela, Odisha, India*
SUN H. PARK • *Department of Cancer Biology and Comprehensive Cancer Center, Wake Forest School of Medicine, Winston-Salem, NC, USA*
SRIMANTA PATRA • *Department of Life Science, National Institute of Technology Rourkela, Rourkela, Odisha, India*
LAURA M. PÉREZ • *Universidad Europea de Madrid, Madrid, Spain; Instituto de Investigación Hospital 12 de Octubre (I+12), Madrid, Spain*
LIUDMILA POLONCHUK • *Roche Pharma Research and Early Development, Roche Innovation Center Basel, F. Hoffmann-La Roche Ltd., Basel, Switzerland*
PRAKASH P. PRAHARAJ • *Department of Life Science, National Institute of Technology Rourkela, Rourkela, Odisha, India*
PRASHANTHI RAMESH • *LEXOR, Center for Experimental and Molecular Medicine, Amsterdam UMC, University of Amsterdam, Cancer Center Amsterdam, Amsterdam, Netherlands; Oncode Institute, Amsterdam, Netherlands*
SARBARI SAHA • *Department of Life Science, National Institute of Technology Rourkela, Rourkela, Odisha, India*
YUSUKE SHIOZAWA • *Department of Cancer Biology and Comprehensive Cancer Center, Wake Forest School of Medicine, Winston-Salem, NC, USA*
EDWARD F. SROUR • *Department of Medicine, Indiana University School of Medicine, Indianapolis, IN, USA; Department of Microbiology and Immunology, Indiana University School of Medicine, Indianapolis, IN, USA*
DAISUKE SUGIYAMA • *Department of Research and Development of Next Generation Medicine, Faculty of Medical Sciences, Kyushu University, Fukuoka, Japan; Center for Clinical and Translational Research, Kyushu University Hospital, Fukuoka, Japan; Incubation Center for Advanced Medical Science, Kyushu University, Fukuoka, Japan; Department of Clinical Study, Center for Advanced Medical Innovation, Kyushu University, Fukuoka, Japan; Kyushu University, Station for Collaborative Research1 4F, Fukuoka, Japan*
BING-DONG SUI • *State Key Laboratory of Military Stomatology, Center for Tissue Engineering, School of Stomatology, Fourth Military Medical University, Xi'an, Shaanxi, China; Research and Development Center for Tissue Engineering, Fourth Military Medical University, Xi'an, Shaanxi, China*
RUSSELL S. TAICHMAN • *Department of Periodontics and Oral Medicine, University of Michigan School of Dentistry, Ann Arbor, MI, USA*
KEAI SINN TAN • *Institute of Biomedical and Pharmaceutical Sciences, Guangdong University of Technology, Guangzhou, China*

Michèle Weiss-Gayet • *Institut NeuroMyoGène, Univ Lyon, Université Claude Bernard Lyon 1, INSERM U1217, CNRS UMR5310, Lyon, France*

Pan Zhao • *State Key Laboratory of Military Stomatology, Center for Tissue Engineering, School of Stomatology, Fourth Military Medical University, Xi'an, Shaanxi, China; Research and Development Center for Tissue Engineering, Fourth Military Medical University, Xi'an, Shaanxi, China*

Bin Zhu • *State Key Laboratory of Military Stomatology, Center for Tissue Engineering, School of Stomatology, Fourth Military Medical University, Xi'an, Shaanxi, China; Research and Development Center for Tissue Engineering, Fourth Military Medical University, Xi'an, Shaanxi, China*

Methods in Molecular Biology (2019) 2002: 1–11
DOI 10.1007/7651_2018_180

Published online: 30 August 2018

In Vitro Maintenance of Multipotent Neural Crest Stem Cells as Crestospheres

Sofie Mohlin and Laura Kerosuo

Abstract

Neural crest cells are a critical source of many cell types of the vertebrate body. However, as a stem cell population they are peculiar because of the transient nature of their stem cell niche; soon after the multipotent neural crest cells are specified in the neuroepithelium, they become mesenchymal cells that migrate into various destinations in early embryos. These rapid in vivo changes during neural crest development complicate the studies on their stem cell properties. Crestospheres are in vitro maintained primary cultures of premigratory neural crest cells that maintain a mixture of neural crest stem and progenitor cells for weeks without spontaneous differentiation, including the multipotent neural crest stem cells. Here, we describe how crestosphere cultures are initiated from either cranial or trunk levels of chick embryos. Alternatively, the same culture conditions can be used to maintain human embryonic stem cell-derived neural crest cells as crestospheres. Thus, crestospheres provide a useful tool for studies on neural crest stemness.

Keywords Chick embryo, Cranial neural crest, Crestospheres, Human ES cell-derived neural crest, Multipotency, Neural crest, Self-renewal, Trunk neural crest

1 Introduction

The neural crest is a transient stem cell population that arises from the developing neural ectoderm at all axial levels in vertebrates. Neural crest cells are multipotent, giving rise to melanocytes as well as peripheral neurons and glia. Additionally, specific regions such as the cephalic, cardiac, or trunk neural crest cells give rise to axial-level specific cell types such as cranial bone and cartilage, the cardiac septum of the outflow tract of the heart, or the endocrine cells of the adrenal medulla. These different axial levels have been shown to be actively maintained by distinct cranial and trunk gene regulatory circuits [1], yet many open questions remain regarding how neural crest stemness is maintained. According to the current understanding, neural crest cells are highly multipotent at the premigratory stage in the dorsal neural tube, and gradually lose their stem cell potential during the course of migration [2], but very little is known about the details of this process (Fig. 1a).

Cell culture models are useful tools to address mechanistic questions about stem cell fate. Primary neural crest cell cultures

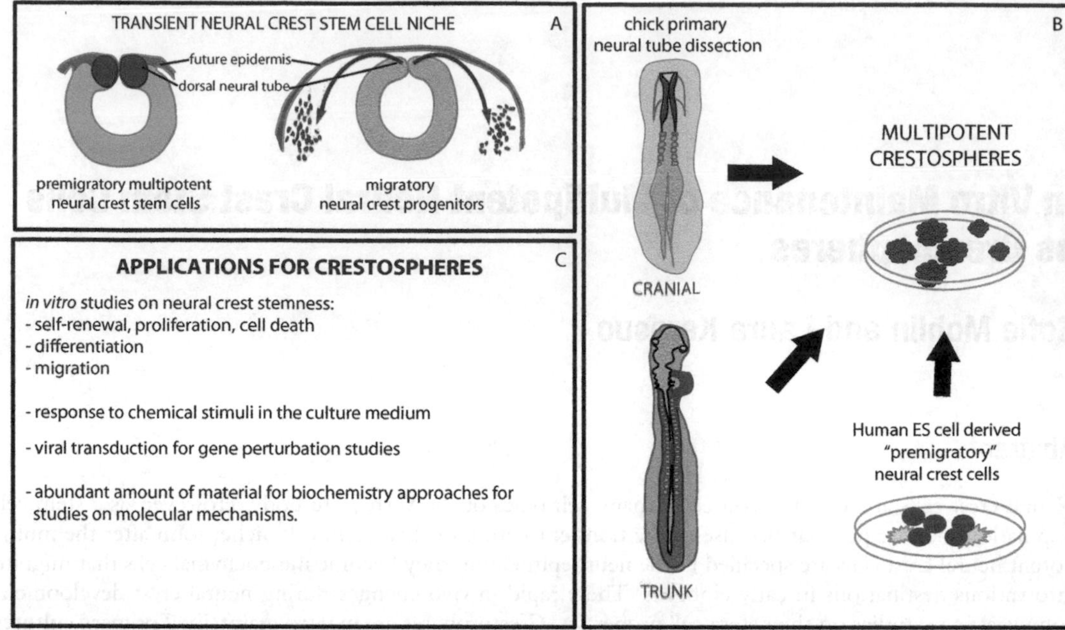

Fig. 1 Crestosphere technique. (**a**) According to the current understanding, neural crest cells are highly multipotent at the premigratory stage of their development immediately after specification at the dorsal neural tube. The transient niche is left empty after the neural crest cells become mesenchymal and migrate to multiple destinations in the vertebrate body and gradually lose their stemness. (**b**) Cartoon visualizing different sources of crestosphere cultures and (**c**) a list of applications (modified from Kerosuo et al. 2015). Reprinted from [3], Copyright 2015, with permission from Elsevier

have been established for decades [4–7] but the cultures have been initiated with migrating neural crest cells under conditions that promote spontaneous differentiation, thus restricting the maintenance of self-renewal and multipotency. Crestosphere cultures "stop time" and maintain neural crest cells in a multipotent stem cell state providing a powerful tool for studies on the characteristics of premigratory neural crest stem cells [2]. They can be utilized for studying self-renewal, proliferation, differentiation, or cell death assays, and their developmental potential can also be addressed by performing transplantation into donor embryos following in vitro manipulation [3, 8]. Crestospheres are well suitable for various molecular biology applications including determination of gene expression, epigenetics, or protein–protein interactions, and to study neural crest cell response to various stimuli induced by either modifications in the cell culture medium or by using gene perturbation techniques.

We have established protocols for both cranial- [3] and trunk-derived crestospheres [9]. In brief, crestospheres are established from dissected and dissociated neural tubes from embryos that are at the premigratory stage of their neural crest development (4–7 somite stage for cranial and 16–20 somite stage for trunk). Pools of premigratory neural crest containing neural tubes from 4 to 6 embryos constitute one crestosphere culture that can then be expanded and kept in culture for several weeks. Crestospheres are cultured as free-floating neuroepithelial 3D clusters of neural crest stem and progenitor cells, similar to neurosphere cultures [10], under minimum adherence and serum-free conditions with a modified protocol for optimal dorsal neural tube/neural crest cell identity and growth. Here, we describe the detailed procedure of successful establishment of crestospheres from both cranial and trunk neural crest (Fig. 1b, c).

2 Materials

Ringer's and PBS solutions can be stored at room temperature. Basic neural crest (NC) medium should be stored at +4 °C and preferably pre-warmed to 37 °C before use.

2.1 Embryos

1. Commercially purchased fertilized chicken eggs
2. Incubator at 37.5 °C/100 °F (*see* **Note 1**)

2.2 Isolation of Neural Tubes

1. Seventy percent EtOH
2. Surgical instruments including forceps, fine scissors, and micro scissors
3. Autoclaved (*optional*) Whatman filter paper squares with punched holes (use a regular paper puncher for office use) in the middle
4. Ringer's solution: [Solution-1: 144 g NaCl, 4.5 g $CaCl\cdot 2H_2O$, 7.4 g KCl, and ddH_2O to 500 ml; Solution-2: 4.35 g $Na_2HPO_4\cdot 7H_2O$, 0.4 g KH_2PO_4, and ddH_2O to 500 ml (adjust final pH to 7.4)] (*see* **Note 2**)
5. Sterile PBS: (NaCl, KCl, Na_2HPO_4, KH_2PO_4, and H_2O)

2.3 Culture

1. Basic NC medium: DMEM with 4.5 g/l glucose, 1% penicillin/streptomycin, chick embryo extract (7.5%), B27 (1×), insulin growth factor –I (IGF-I, 20 ng/ml), and basic fibroblast growth factor (bFGF, 20 ng/ml) (*see* **Note 3**)
2. (*Optional*) In-house produced chick embryo extract: 11-day--old chick embryos, DMEM, gauze, 10 ml syringe, hyaluronidase, and ultracentrifuge (*see* **Note 4**)

3. Additions to NC medium—60 nM (cranial) or 180 nM (trunk) retinoic acid (*see* **Note 5**) and 25 ng/ml (trunk) bone morphogenetic protein (BMP)-4
4. Ultralow attachment T25 culture flasks (*see* **Note 6**)
5. Sterile-filtered Accutase® solution for cell culture

3 Methods

All work should be carried out in a clean environment. Dissociation of dissected neural tubes and all work thereafter should be carried out in sterile cell culture hood.

3.1 Isolation of Neural Tubes

1. Incubate fertilized eggs to 4–8 somite stage (Hamburger Hamilton HH stage 8-/9- for cranial crestospheres) or 16–20 somite stage embryos (HH stage 13-/14- for trunk crestospheres)
2. Spray eggs with 70% EtOH
3. Open eggs by cracking the shell from 1/3 below the top by using robust forceps. Pour out albumin through this hole and use forceps to pull out thick albumin. Gradually, remove all albumin and gently wipe off remaining thick albumin surrounding the embryo using Kim Wipes tissue paper. If the embryo is on the side, gently roll the yolk by using the robust forceps (importantly avoid touching the embryo). Collect the embryo by placing a squared 1 cm × 1 cm filter paper on top of the embryo so that the embryo is within the stenciled non-paper area in the middle, and the vitelline membrane will then be attached to the paper. Cut off the filter paper from the yolk and transfer the embryo to Ringer's solution in a petri dish.
4. By careful dissection under a microscope, isolate neural tubes in Ringer's solution from neighboring tissue using micro scissors. Start by placing the embryo ventral side up and cut the endoderm from the middle as if you were "opening a jacket." Now, carefully cut out the neural tube, and importantly exclude all neighboring mesoderm and the notochord.
5. For cranial neural tubes, exclude the very anterior tip (that does not produce neural crest), and collect the neural tube up to second (2nd) somite level. For trunk, aim for the neural tube between the tenth and fifteenth (10–15) somite level. Each established culture should consist of 4–6 pooled neural tubes (*see* **Note 7**).
6. Mechanically dissociate the complete pool of neural tubes by pipetting up and down 30 times in 50 μl Ringer's solution or 1× PBS using a p200 tip in an Eppendorf tube. Monitor under

a microscope that the tissue has been dissociated into small (20–50 cell) clumps.

7. Prepare basic NC medium in an ultralow attachment T25 flask and add retinoic acid (cranial and trunk) and BMP-4 (trunk) appropriate to total volume (concentration of retinoic acid dependent on axial-level source of neural tubes) (*see* **Note 8**). Transfer dissociated tissue pieces to 1–2 ml prepared NC medium supplemented with retinoic acid (cranial and trunk) and BMP-4 (trunk) (*see* **Note 9**).
8. When culturing human embryonic stem (ES) cell-derived crestospheres, induce neural crest development according to previously published protocol [11]. At day 7–9 in neural crest-inducing medium, transfer the neuroepithelial spheres to NC medium in ultralow attachment T25 flasks.

3.2 Crestosphere Cultures

1. (*Optional*) Prepare in-house produced chick embryo extract by rinsing headless 11-day-old chick embryos with cold DMEM on a double layer of gauze on a 500-ml beaker until blood is removed. Transfer embryos to 10 ml syringe and push through into a 50-ml Falcon tube. Weigh embryos and dilute in DMEM (1 ml DMEM/ 1 g of minced embryos) and stir at +4 °C overnight. Add ice chilled hyaluronidase (4×10^{-5} g/1 g of minced embryos) and stir at +4 °C for 1 h. Ultracentrifuge lysates for 30 min at 46,000 × *g* and filter sterilize (0.45 μm) the clear supernatant. Aliquot the chick embryo extract (~5–10 ml per tube) and store at −80 °C (*see* **Note 4**).
2. Due to the small starting volume, culture the dissociated tissue pieces in T25 flasks placed in upright position the first 2–3 days (*see* **Note 10**).
3. After 2–3 days, add another 1–2 ml of basic NC medium and add RA corresponding to new volume ($V = 3$–4 ml) (*see* **Note 11**). Always use a stripette to measure the current culture volume. In addition, upon adding fresh medium, slightly dissociate the spheres by pipetting them roughly ten times up and down towards the wall of the culture flask. Continue incubation with the T25 flasks in flat ("lying") position from hereafter (critically after a total volume of 3 ml).
4. Following another 2–3 days of incubation, add 3 ml of basic NC medium and RA to new volume ($V = 6$–7 ml) accordingly.
5. Repeat **step 4** following another 2–3 days of culture to reach total culture volume $V = 10$ ml. The user should strive to maintain this culture volume for continuous culturing (*see* **Note 11**).
6. Add fresh retinoic acid (cranial and trunk) and BMP-4 (trunk) every 2 to 3 days (*see* **Note 12**).

7. Crestospheres can be kept in culture for several weeks with maintained self-renewal capacity and expression of neural crest-associated genes (*see* **Note 13**).

3.3 Validation and Applications

1. Before starting experiments with the newly established crestospheres, the end user should always validate the success of the procedure by assessing the expression of neural crest genes using in situ hybridization, Q-PCR, and/or immunostaining (*see* **Note 14**). Successful cultures are highly enriched in premigratory neural crest cells (expressing, e.g., FoxD3), but markers for more differentiated neural crest progenitors (e.g., p75, and HNK1), or low levels of central nervous system (CNS) neural genes (e.g., Sox2) may also be present (Fig. 2).
2. Crestospheres provide a useful tool to study neural crest stemness and differentiation mechanisms and can be used for a myriad of in vitro applications, such as assays to validate cell proliferation, apoptosis, or migratory behavior (Figs. 1 and 2). Crestospheres are highly suitable for cell extraction for downstream gene expression studies, epigenetics, or protein–protein interactions. Gene expression in crestospheres can be modified by using retroviral transduction or chemical activators/inhibitors added to the culture medium. The developmental potential of crestospheres can also be tested in vivo by transplanting the spheres back into donor embryos to join the migratory neural crest stream (Fig. 2, and *see* **Note 15**). Single-cell level analysis can be performed either by using confocal imaging or by imaging cryosections (*see* **Note 16** for embedding and **Note 17** for whole mount imaging).

4 Notes

1. Chick embryo development is monitored according to Hamburger Hamilton (HH) staging system, once determined by incubation at 37 °C. This temperature might need to be adjusted by the end user.
2. Ringer's solution: Dilute 40× Solution-1 and 40× Solution-2 to 1× in ddH_2O. Add the solutions separately to water to avoid precipitation of calcium. Adjust pH to 7.4 (add ~1–2 drops of 1 M NaOH) and filter sterilize using 0.22-μm bottle top filters into autoclaved flasks.
3. Prepare basic NC medium in advance, facilitating transfer of dissociated neural tubes to culture without delay in time. Basic NC media is stable for ~1 week at 4 °C.
4. Chick embryo extract (CEE) can be produced by the end user or bought commercially. In both cases, run the extract through

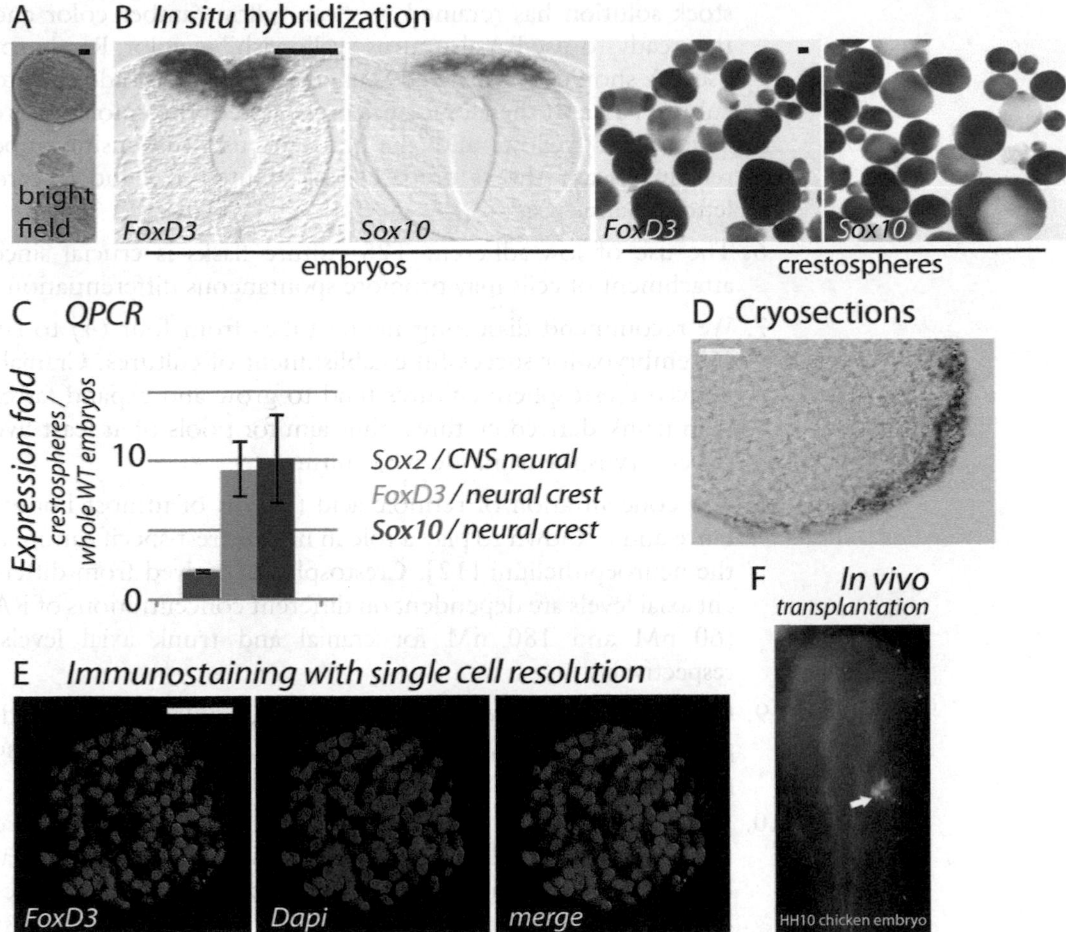

Fig. 2 Examples of techniques used for crestosphere validation and functional studies. (**a**) Bright field microscopy. (**b**) Crestospheres are highly enriched for neural crest cells as shown by in situ hybridization of *FoxD3* and *Sox10* expression in premigratory neural crest in the chicken embryonic dorsal neural tube and in crestospheres (**c**) or by QPCR showing high *FOXD3* and *SOX10*, whereas the expression levels of the neural stem cell gene *SOX2* are low. (**d**) Cryosections of crestospheres provide a tool for in-depth examination of in situ hybridization, as exemplified by the expression of the *EPAS1* gene, verifying that RNA expression is only localized at the outer edge of the sphere. (**e**) Confocal high-resolution imaging of a whole mount immunostained crestosphere by using an antibody to FoxD3. (**f**) In vivo transplantation showing crestospheres after implantation into donor embryo at HH10. Scale bars represent 50 μm. (**a–c** and **e**, **f**: Reprinted from [3], Copyright 2015, with permission from Elsevier)

a 0.22-μm bottle top filter, and store in 1–5 ml aliquots at −20 °C or −80 °C.

5. Retinoic acid (RA/ATRA) is resuspended in DMSO as a stock solution, e.g., in 50 μl aliquots in −20 °C. Store and handle RA protected from light. Before use, thaw the aliquot in room temperature to prepare a ready-to-use dilution in PBS/culture medium, and importantly ensure at each occasion that the RA

stock solution has retained its clear yellow/amber color and that ready-to-use RA also turns "yellowish" in color. Ready-to-use RA should be vortexed immediately prior to addition to culture flasks. If the aliquot has precipitated, the color is more "neon-like" yellow, and the ready-to use suspension tube remains clear. Always prepare fresh diluted RA and discard leftovers after use.

6. The use of low-adherent T25 culture flasks is crucial since attachment of cells may promote spontaneous differentiation.
7. We recommend dissecting neural tubes from four (4) to six (6) embryos for successful establishment of cultures. Cranial-derived crestosphere cultures tend to grow and expand faster than trunk-derived cultures, thus aim for pools of at least five (5) embryos for trunk-derived cultures.
8. The concentration of retinoic acid (RA) is of utmost importance and is known to play a role in neural crest specification in the neuroepithelium [12]. Crestospheres derived from different axial levels are dependent on different concentrations of RA (60 nM and 180 nM for cranial and trunk axial levels, respectively).
9. Once neural tubes from all embryos to be pooled are dissected, proceed immediately to dissociation of sample as delays in the process may cause cell death and tissue degradation.
10. For successful early expansion of crestosphere cultures, we recommend a small culture volume ($V = 2$ ml for cranial cultures with six embryos, and 1–2 ml for trunk) and consequently incubating the T25 flask in upright position the first days following neural tube isolation. A small starting volume is critical for the success of the culture, too dilute of a cell concentration may kill off the culture.
11. For gentle and stepwise expansion of crestospheres, small volumes of basic NC medium are added in the beginning of crestosphere culturing. This is continued until total culture volume reaches 10 ml. A good rule of thumb is to add 25–50% more medium when you can clearly see an increase in crestosphere number by eye. A full flask with 10 ml medium should contain hundreds of crestospheres. Monitor the cells daily, and add new medium more frequently, even daily, if the cultures grow fast, but as in the beginning, avoid too dilute conditions.
12. Add new retinoic acid (RA) and BMP-4 every 2–3 days due to high turnover rates. Measure total culture volume by using a stripette and add factors accordingly.

13. Crestospheres can be kept in culture for several weeks with maintained expression of neural crest-associated genes and self-renewal capacity. However, the longest time point examined is 7 weeks and at this time point the proliferation capacity is significantly decreased [3].

14. Briefly, QPCR is performed by collecting 1–2 ml of sphere cultures in Eppendorf tubes by centrifugation at maximum speed for 1 min at +4 °C. Remove supernatant and store pellets at −80 °C until RNA extraction (method of choice by end user) and further QPCR analysis (method of choice by end user). Alternatively, allow crestospheres to sink down to the bottom of the tube (5–10 min), gently remove all culture medium, and replace with respective RNA, DNA, or protein extraction lysis buffer.

    Crestospheres can also be assessed for in situ hybridization according to protocols for whole embryos [13, 14]. Unlike floating chicken embryos, crestospheres sink to the bottom of the tube during the hybridization step and washes, making the procedure straightforward. In order to not lose any crestospheres during washes, always let crestospheres sink to the bottom of the tube (5–10 min), and leave ~100 μl of the previous wash on the bottom of the tube before adding 1 ml of the new wash.

15. Transgenic GFP-positive chickens are available from Clemson Public Services Activities, Clemson University, SC, USA and are useful for in vivo transplantation applications or for producing chimeras for clonal studies. Crestospheres can also be transduced by using lentiviruses, with a typical transfection success rate of ~50%.

16. Wild type, in situ hybridized, or immunostained crestospheres can be embedded for cryosectioning. Briefly, crestosphere cultures are prepared by sucrose gradient (5% sucrose for 10 min followed by 15% sucrose for 2–4 h at room temperature, allow spheres to collect at bottom of tube). Crestospheres are further primed in gelatin overnight at 37 °C. Embed crestospheres in gelatin using embedding molds. Make sure that the crestospheres do not sink to the bottom of the mold, but rather position in the middle seen from all axes (position can be adjusted as gelatin is solidifying). Stiffened gelatin molds are snap-frozen by dipping the entire mold briefly but repeatedly in liquid nitrogen until complete samples are frozen. Store at −20 °C for 5 min before transferring samples to Eppendorf tubes for long-term storage at −80 °C before cryosectioning. Thickness for cryosections for immunostaining should range from 10 to 12 μm, and 15 to 20 μm for crestospheres after whole mount in situ hybridization (in order to get a strong signal).

17. Crestospheres can be imaged whole mount or as cryosections by using regular microscopy (after in situ hybridization) or confocal microscopy (fluorescent signal). *For high-resolution imaging*, use either of the following procedures: (a) prepare a "chamber" for the floating spheres in PBS by applying a Vaseline wall (inject it from a 3-ml syringe through a $25G^{5/8}$ needle) to make a small rectangle on a glass cover slip, pipette the spheres into the chamber, and cover it by using another glass cover slip; (b) transfer crestospheres to cavity slides and monitor for quicker but less detailed analysis. *For lower magnification bright field imaging* of pools of crestospheres after in situ hybridization, add a ~0.5-cm thick agarose (1% in H_2O) layer on the bottom of the wells in a 24-well plate and let solidify. Place crestospheres in PBS-0.2% Tween on top of the agarose bed and image by using a dissection scope with a camera attached. Even the lightning by using a light source with adjustable, flexible arms.

Acknowledgments

This work was funded by the Swedish Childhood Cancer Foundation, Thelma Zoéga's Foundation, Hans von Kantzow's Foundation, the Mary Béve Foundation, the Royal Physiographic Society of Lund, Magnus Bergvalls Stiftelse (to S.M.) and the Academy of Finland, Sigrid Juselius Foundation, and Ella and Georg Ehrnrooth's Foundation (to L.K.), and in part by the Division of Intramural Research of the National Institute of Dental and Craniofacial Research at the National Institutes of Health, Department of Health and Human Services.

References

1. Simoes-Costa M, Bronner ME (2016) Reprogramming of avian neural crest axial identity and cell fate. Science 352(6293):1570–1573. https://doi.org/10.1126/science.aaf2729
2. Dupin E, Calloni GW, Coelho-Aguiar JM, Le Douarin NM (2018) The issue of the multipotency of the neural crest cells. Dev Biol. https://doi.org/10.1016/j.ydbio.2018.03.024
3. Kerosuo L, Nie S, Bajpai R, Bronner ME (2015) Crestospheres: long-term maintenance of multipotent, premigratory neural crest stem cells. Stem Cell Reports 5(4):499–507. https://doi.org/10.1016/j.stemcr.2015.08.017
4. Cohen AM, Konigsberg IR (1975) A clonal approach to the problem of neural crest determination. Dev Biol 46(2):262–280
5. Sieber-Blum M, Cohen AM (1980) Clonal analysis of quail neural crest cells: they are pluripotent and differentiate in vitro in the absence of noncrest cells. Dev Biol 80(1):96–106
6. Lahav R, Ziller C, Dupin E, Le Douarin NM (1996) Endothelin 3 promotes neural crest cell proliferation and mediates a vast increase in melanocyte number in culture. Proc Natl Acad Sci U S A 93(9):3892–3897
7. Trentin A, Glavieux Pardanaud C, Le Douarin N, Dupin E (2004) Self-renewal capacity is a widespread property of various types of neural crest precursor cells. Proc Natl Acad Sci U S A 101(13):4495–4500
8. Kerosuo L, Bronner ME (2016) cMyc regulates the size of the premigratory neural crest

stem cell pool. Cell Rep 17(10):2648–2659. https://doi.org/10.1016/j.celrep.2016.11.025

9. Mohlin S, Kunttas E, Persson CU, Abdel-Haq R, Castillo A, Murko C, Bronner ME, Kerusou L (2018) Maintaining trunk neural crest cells as crestospheres. bioRxiv. https://doi.org/10.1101/391599

10. Reynolds BA, Weiss S (1992) Generation of neurons and astrocytes from isolated cells of the adult mammalian central nervous system. Science 255(5052):1707–1710

11. Bajpai R, Chen D, Rada Iglesias A, Zhang J, Xiong Y, Helms J, Chang C-P, Zhao Y, Swigut T, Wysocka J (2010) CHD7 cooperates with PBAF to control multipotent neural crest formation. Nature 463(7283):958–962

12. Martinez-Morales P, Diez Del Corral R, Olivera-Martnez I, Quiroga A, Das R, Barbas J, Storey K, Morales A (2011) FGF and retinoic acid activity gradients control the timing of neural crest cell emigration in the trunk. J Cell Biol 194(3):489–503

13. Kerosuo L, Bronner M (2014) Biphasic influence of Miz1 on neural crest development by regulating cell survival and apical adhesion complex formation in the developing neural tube. Mol Biol Cell 25(3):347–355

14. Acloque H, Wilkinson DG, Nieto MA (2008) Chapter 9 in situ hybridization analysis of chick embryos in whole-mount and tissue sections. In: Marianne B-F (ed) Methods in cell biology, vol 87. Academic, New York, pp 169–185. https://doi.org/10.1016/S0091-679X(08)00209-4

Methods in Molecular Biology (2019) 2002: 13–27
DOI 10.1007/7651_2018_176

Published online: 23 January 2019

Analysis of Hematopoietic Niche in the Mouse Embryo

Keai Sinn Tan, Nathalie Brouard, and Daisuke Sugiyama

Abstract

The development, differentiation, and maturation of hematopoietic cells are regulated by the intrinsic and extrinsic regulation. Intrinsic activity is affected by cell autonomous gene expression and extrinsic factors originate from the so-called niche surrounding the hematopoietic cells. It remains unclear why the hematopoietic sites are shifted during embryogenesis. Flow cytometry and immunohistochemistry enable us to study embryonic regulation of hematopoietic niche in the mouse embryo.

Keywords Embryo, Hematopoietic, Mouse, Niche

1 Introduction

Hematopoiesis is a complex process that takes place at several anatomical locations both intra- and extraembryonically. In the mouse, the sites of hematopoiesis change throughout approximately of a 20-day gestation period. Hematopoiesis starts with the yolk sac at 7.5 days post-coitum (dpc) where primitive erythroid cells form, followed by erythroid progenitor cells at approximately 8.25 dpc. The erythroid progenitor cells seed the fetal liver at 12.5 dpc. This primitive hematopoiesis changes to definitive hematopoiesis and begins to sustain the adult blood system through hematopoietic stem cells. Up to date, there is controversy about where exactly hematopoietic stem cells are generated; be it sites in the extraembryonic mesoderm or the intraembryonic para-aortic splanchnopleural (p-Sp) mesoderm/aorta-gonad-mesonephros (AGM). Fetal liver is a hematopoietic organ where the primitive hematopoiesis draws to close and definitive hematopoiesis takes over, leading to it being the major site for the expansion and differentiation of hematopoietic stem cells before migration to the muscle and bone marrow; and that fetal spleen is a hematopoietic organ which likely fills the "gap" between fetal liver and bone marrow hematopoiesis. During embryogenesis, hematopoietic cell fate is determined by two types of activities: (1) intrinsic regulation, programmed primarily by cell autonomous gene expression; and (2) extrinsic factors, which originate from the so-called niche cells surrounding hematopoietic cells [1, 2]. Adhesion receptors are expressed by hematopoietic stem and progenitor cells and that provide specific cell–cell interactions for their establishment and self-renewal ability [3]. However, much effort remains to be carried out to identify hematopoietic niches as well as those signals that could dictate hematopoietic cell development at specific stages.

Previously, we reported that endothelial cells of the placenta regulate hematopoietic cell clusters through SCF/Kit regulation in embryonic mouse [4]. The mesonephric cells of the AGM region of a mouse embryo expressed CSF1 and that intra-aortic clusters expressed CSF1 receptor, in which the CSF1/CSF1R signalling modulates intra-aortic cluster myeloid differentiation through *Cebpα*, *Ifr8*, *Cebpε*, and *Gfi1* [5]. In fetal liver, Dlk-1-expressing hepatoblasts primarily expressed EPO and SCF, genes encoding erythropoietic cytokines. In *Map2k4*$^{-/-}$ mouse embryos which lack hepatoblasts, the cell numbers were decreased, suggesting that hepatoblasts comprise a niche for erythropoiesis through cytokine secretion [6]. In addition, interaction of Integrin beta-1 with extracellular matrix promotes adherence of hematopoietic stem cells to fetal liver and that hepatoblasts expressed extracellular matrix under the control of TGF-beta-1 in the fetal liver [7]. A population of cells lacking colony forming unit fibroblast activity and exhibiting hepatocyte progenitor profile reportedly support megakaryocyte production in 13.5 dpc mouse liver [8]. Hepatic stellate cells at 12.5, 14.5, and 16.5 dpc expressed *Igf2*. Also, the hepatic stellate cells highly express *fibronectin* at 12.5 dpc and *vitronectin* at 14.5 and 16.5 dpc [9]. In the spleen, endothelial cells and mesenchymal-like cells expressed SCF and IGF-1 and that cytokines regulate the erythropoiesis [10]. Before homing to bone marrow, the fetal liver hematopoietic progenitor cells reside in the muscle tissue [11].

To identify and isolate specific cell compartments, we utilize approaches such as the flow cytometry and immunohistochemistry to identify those cells surrounding hematopoietic cells in hematopoietic organs of the mouse embryo. This approach enables us to identify niche cells and study the extrinsic regulation of those cells on the development, differentiation, and maturation of those hematopoietic cells.

2 Materials

2.1 Flow Cytometry

A. p-Sp/AGM region and placenta

1. Mouse embryos at 9.5 dpc (18-22 SPs), 10.5 dpc (30-34 SPs), and 11.5 dpc (42-46 SPs)
2. 21-Gauge needle (Terumo)
3. Collagenase solution: 1 mg/ml collagenase in medium supplemented with 10% fetal bovine serum

4. 40-μm Cell strainer (BD Pharmingen)
5. Lysing buffer: BD Pharm Lyse lysing buffer (BD Pharmingen)
6. Propidium iodide (PI) buffer (Invitrogen)
7. The following antibodies are used to isolate intra-aortic cluster: PE- or PE-Cy7-conjugated anti-mouse CD31 (BD Biosciences), PE-Cy7- or APC-Cy7-conjugated anti-mouse c-Kit, and Pacific Blue-conjugated anti-mouse CD34 (eBioscience, Santa Clara, CA). The following antibodies are used to isolate endothelial and mesenchymal cell populations in placenta: FITC-conjugated anti-mouse Ter119 (eBioscience), PE-conjugated anti-mouse CD31 (BD Biosciences), APC-conjugated anti-mouse c-Kit (BD Biosciences), PE-Cy7-conjugated anti-mouse CD45 (BioLegend), and Pacific Blue-conjugated anti-mouse CD34 (eBioscience)
8. Flow cytometry: FACSAria cell sorter (BDIS)
9. RNAlater (Ambion)

B. Placenta

1. Mouse embryos at 10.5 dpc, 11.5 dpc, and 12.5 dpc
2. 21-Gauge needle (Terumo)
3. Collagenase solution: 1 mg/ml collagenase in medium supplemented with 10% fetal bovine serum
4. 40-μm Cell strainer (BD Pharmingen)
5. Lysing buffer: BD Pharm Lyse lysing buffer (BD Pharmingen)
6. Propidium iodide (PI) buffer (Invitrogen)
7. The following antibodies are used to isolate intra-aortic cluster: PE- or PE-Cy7-conjugated anti-mouse CD31 (BD Biosciences), PE-Cy7- or APC-Cy7-conjugated anti-mouse c-Kit, and Pacific Blue-conjugated anti-mouse CD34 (eBioscience, Santa Clara, CA). The following antibodies are used to isolate endothelial and mesenchymal cell populations in placenta: FITC-conjugated anti-mouse Ter119 (eBioscience), PE-conjugated anti-mouse CD31 (BD Biosciences), APC-conjugated anti-mouse c-Kit (BD Biosciences), PE-Cy7-conjugated anti-mouse CD45 (BioLegend), and Pacific Blue-conjugated anti-mouse CD34 (eBioscience)
8. Flow cytometry: FACSAria cell sorter (BDIS)
9. RNAlater (Ambion)

C. Fetal liver

1. Mouse embryos at 11.5 dpc, 12.5 dpc, 13.5 dpc, 14.5 dpc, 16.5 dpc, and 19.5 dpc
2. Scalpel
3. Collagenase type I (Worthington ref. X4J7447) (3 mg/ml) in PBS
4. 45–70-μm Cell strainers (BD Pharmingen)
5. 4′,6-Diamidino-2-phenylindole (DAPI) (Sigma-Aldrich) or propidium iodide (PI) buffer (Invitrogen)
6. The following antibodies are used to isolate endothelial and mesenchymal cell populations in fetal liver: biotin-conjugated anti-mouse CD45 (BD Biosciences), biotin-conjugated anti-mouse Ter119 (BD Bioscience), PE-Cy7-conjugated anti-mouse VCAM-1 (BioLegend), PE-conjugated anti-mouse CD51 (eBioscience), FITC-conjugated anti-mouse CD31 (BD Biosciences), APC-conjugated anti-mouse PDGFRα (eBioscience), and a secondary antibody Streptavidin-conjugated anti-mouse APC-Cy7 (BioLegend)
7. Flow cytometry: FACSAria cell sorter (BDIS)/FortessaX-20 SORP (BDIS)
8. RNAlater (Ambion)

D. Fetal muscle

1. Mouse embryos at 14.5 dpc, 15.5 dpc, 16.5 dpc, 17.5 dpc, 18.5 dpc, and 19.5 dpc
2. 29–32-Gauge needle (Terumo)
3. Collagenase solution: 3 mg/ml collagenase in medium supplemented with 10% fetal bovine serum
4. 70-μm Cell strainer (BD Pharmingen)
5. Propidium iodide (PI) buffer (Invitrogen)
6. The following antibodies are used to isolate study muscle HSPC populations: PE-conjugated anti-mouse CD45 antibody, PE-Cy7-conjugated anti-mouse c-Kit, Pacific Blue-conjugated anti-mouse F4/80, FITC-conjugated anti-mouse Sca-1 (all from BioLegend), and anti-mouse CD16/32 Fc binding blocker (eBioscience)
7. Flow cytometry: FACSAria cell sorter (BDIS)
8. RNAlater (Ambion)

E. Fetal spleen

1. Mouse embryos at 16.5 dpc and 19.5 dpc
2. Collagenase solution: 3 mg/ml collagenase (Worthington Biochemical Corporation) in medium supplemented with 10% fetal bovine serum
3. 70-μm Cell strainer (BD Pharmingen)
4. Propidium iodide (PI) buffer (Invitrogen)
5. The following antibodies are used to isolate endothelial cells and mesenchymal-like cells in the spleen: Pacific Blue-, APC-, and PE-Cy7-conjugated anti-mouse CD45 (BioLegend); PE-Cy7-, APC-Cy7-, and APC-conjugated anti-mouse Ter119 (eBioscience); PE- and APC-conjugated anti-mouse CD31 (BD Bioscience); APC-conjugated anti-mouse LYVE-1 (MBL); and FITC-conjugated anti-mouse DLK-1 (MBL)
6. Flow cytometry: FACSAria cell sorter (BDIS)
7. RNAlater (Ambion)

2.2 Immunohistochemistry

A. p-Sp/AGM region

1. Mouse embryos at 10.5 dpc
2. 20-μm tissue slice
3. Primary antibody solution for the AGM region: PBS containing 1% BSA with appropriate dilutions of the following primary antibodies: goat anti-mouse c-Kit (R&D Systems), rabbit anti-mouse CSF1R (Novus Biologicals), anti-mouse Thrombopoietin receptor antibody (Kyowa Hakko Kirin Co.), and rabbit anti-mouse Cleaved Caspase-3 (Cell Signaling Technology)
4. Secondary antibodies in PBS containing 1% BSA: Alexa Fluor 488 donkey anti-rabbit IgG (Invitrogen), Alexa Fluor 546 donkey anti-goat IgG (Invitrogen), and Alexa Fluor 568 donkey anti-goat IgG (Invitrogen), with TOTO-3 iodide (Invitrogen)
5. Fixative: 2% paraformaldehyde in PBS
6. 30% Sucrose in PBS
7. Liquid nitrogen
8. Blocking buffer: 1% BSA in PBS
9. OCT compound (SAKURA)
10. Cryostat: CM1900 UV (Leica)
11. Glass slides (Matsunami)
12. Coverslips
13. Tyramide Signal Amplification System (PerkinElmer)
14. Fluorescent mounting medium (Dako Corporation)

15. Confocal microscope: FluoView 1000 Confocal Microscope (Olympus)

B. Placenta
1. Mouse embryos at 10.5 dpc and 11.5 dpc
2. 20-μm tissue slice
3. Primary antibody solution for placenta: PBS containing 1% BSA with appropriate dilutions of the following primary antibodies: goat anti-mouse Kit (R&D Systems), rat anti-mouse CD31 (BD Biosciences), rat anti-mouse CD34 (BD Biosciences), rat anti-mouse CD41 (BioLegend), rat anti-mouse CD45 (BioLegend), and rat anti-mouse F4/80 (BioLegend)
4. Secondary antibodies in PBS containing 1% BSA: Alexa Fluor 488 donkey anti-rat IgG (Invitrogen) and Alexa Fluor 568 donkey anti-goat IgG (Invitrogen), as well as TOTO-3 iodide (Invitrogen)
5. Fixative: 2% paraformaldehyde in PBS
6. 30% Sucrose in PBS
7. Liquid nitrogen
8. Blocking buffer: 1% BSA in PBS
9. OCT compound (SAKURA)
10. Cryostat: CM1900 UV (Leica)
11. Glass slides (Matsunami)
12. Coverslips
13. Tyramide Signal Amplification System (PerkinElmer)
14. Fluorescent mounting medium (Dako Corporation)
15. Confocal microscope: FluoView 1000 Confocal Microscope (Olympus)

C. Fetal liver
1. Mouse embryos at 12.5 dpc, 13.5 dpc, and 14.5 dpc
2. 5–8-μm or 20-μm tissue slice
3. Primary antibody solution for the fetal liver: purified rabbit monoclonal anti-mouse VCAM1 (Epitomics), purified anti-mouse CD140a (eBioscience), and purified anti-mouse CD16/32 (BioLegend) in blocking buffer or PBS containing 1% BSA with appropriate dilutions of the following primary antibodies: anti-mouse Dlk-1 (MBL), anti-mouse Lyve-1 (MBL), anti-mouse c-Kit (R&D Systems), anti-mouse SCF (Santa Cruz Biotechnology), anti-mouse

EPO (Santa Cruz Biotechnology), and anti-mouse Ki-67 (Dako Corporation)

4. Secondary antibodies in PBS containing 1% BSA: Alexa Fluor 647 anti-mouse CD150 (BioLegend), Alexa Fluor 555 anti-human/mouse CD42c (produced and conjugated in house), Alexa Fluor 488 donkey anti-rabbit IgG (Jackson ImmunoResearch Laboratories), Alexa Fluor 594 donkey anti-rat IgG (Jackson ImmunoResearch Laboratories), Alexa Fluor 555 donkey anti-rabbit IgG, Alexa Fluor 488 donkey anti-goat IgG, Alexa Fluor 488 donkey anti-rat IgG and Alexa Fluor 546 donkey anti-rat IgG, Alexa Fluor 546 Streptavidin as well as TOTO-3 iodide (Invitrogen)
5. Fixative: 2% or 4% paraformaldehyde in PBS
6. 30% Sucrose in PBS
7. Liquid nitrogen
8. Blocking buffer: 5% BSA, skim milk powder, 0.05% Triton-X100 in 4× SSC supplemented with 2% normal donkey serum (Jackson ImmunoResearch Laboratories) or 1% BSA in PBS
9. OCT compound (SAKURA, Tokyo, Japan)
10. Cryostat: CM1900 UV (Leica)/Microm HM525 Cryostat (Microm)
11. Glass slides (Matsunami)
12. Coverslips
13. Tyramide Signal Amplification System (PerkinElmer)
14. Fluorescent mounting medium (Dako Corporation)
15. Confocal microscope: FluoView 1000 Confocal Microscope (Olympus)

D. Fetal muscle

1. Mouse embryos at 14.5 dpc, 15.5 dpc, 16.5 dpc, 17.5 dpc, 18.5 dpc, and 19.5 dpc
2. Primary antibody solution for the fetal muscle: PBS containing 1% BSA with appropriate dilutions of the following primary antibodies: goat anti-mSCF R/c-Kit (R&D systems), hamster anti-mouse CD31 (Chemicon International), and rat anti-mouse CD45 (BioLegend)
3. Secondary antibodies in PBS containing 1% BSA: Alexa Fluor 488 donkey anti-goat IgG (Invitrogen), Dylight549 donkey anti-rat IgG (Jackson ImmunoResearch Laboratories) and TOTO-3 iodide (Invitrogen), or with Alexa Fluor 633 donkey anti-goat IgG (Invitrogen), DyLight

549 rabbit anti-hamster IgG (Jackson ImmunoResearch Laboratories), Alexa Fluor 488 donkey anti-rat IgG and/or DAPI (Invitrogen)

4. Fixative: 2% paraformaldehyde in PBS
5. 30% Sucrose in PBS
6. Blocking buffer: 1% BSA in PBS
7. Liquid nitrogen
8. OCT compound (SAKURA)
9. Cryostat: CM1900 UV (Leica)
10. Glass slides (Matsunami)
11. Coverslips
12. Tyramide Signal Amplification System (PerkinElmer)
13. Fluorescent mounting medium (Dako Corporation)
14. Confocal microscope: FluoView 1000 Confocal Microscope (Olympus)

E. Fetal spleen

1. Mouse embryos at 16.5 dpc
2. Primary antibody solution for the fetal spleen: PBS containing 1% BSA with appropriate dilutions of the following primary antibodies: rat anti-mouse CD31 (BD Biosciences), rat anti-mouse CD51 (BD Biosciences), rat anti-mouse LYVE-1 (MBL), and rat anti-mouse DLK-1 (MBL)
3. After three PBS washes, incubate sections with appropriate dilutions of the following secondary antibodies in PBS containing 1% BSA: Alexa Fluor 488 donkey anti-rat IgG (Invitrogen), HRP donkey anti-goat IgG (R&D systems), Alexa Fluor 546 Streptavidin (Invitrogen) as well as TOTO-3 iodide (Invitrogen)
4. Fixative: 2% paraformaldehyde in PBS
5. 30% Sucrose in PBS
6. Liquid nitrogen
7. Blocking buffer: 1% BSA in PBS
8. OCT compound (SAKURA)
9. Cryostat: CM1900 UV (Leica)
10. Glass slides (Matsunami)
11. Coverslips
12. Tyramide Signal Amplification System (PerkinElmer)
13. Fluorescent mounting medium (Dako Corporation)

14. Confocal microscope: FluoView 1000 Confocal Microscope (Olympus)

3 Methods

3.1 Flow Cytometry

A. p-Sp/AGM region

Carry out all the procedures at room temperature unless otherwise specified.

1. Obtain the caudal portion of embryos containing the p-Sp/AGM region from pregnant mice (*see* **Note 1**). Count the number of somite pairs (SPs), and dissect the embryos 30–34 SPs (10.5 dpc) for analysis.
2. Tissues were incubated with collagenase solution for 30 min at 37 °C. After incubation, pipette tissue up and down gently for approximately 10 times to obtain a single-cell suspension (*see* **Note 2**).
3. Filter using 40-μm nylon cell strainers.
4. Add PBS to the suspension and centrifuge at 900 × *g* for 5 min.
5. Resuspend the cell pellet in 100–200 μl PBS.
6. Add all color-conjugated antibodies to the single-cell suspension. Add 0.3 μl of each antibody per 1.0 × 10^6 cells.
7. Incubate samples on ice for at least 30 min (*see* **Note 3**).
8. Add PI buffer to the suspension to remove death cells by flow cytometer.
9. Intra-aortic clusters are defined as CD31+/CD34+/Kit+. Set the gate on flow cytometer for each population and begin isolation.
10. Collect isolated cells in RNA later for RNA extraction.

B. Placenta

Carry out all the procedures at room temperature unless otherwise specified.

1. Obtain placentas from pregnant mother. Remove deciduae and umbilical vessels from the placenta (*see* **Note 1**).
2. Pass placenta through 21-gauge needles to disrupt tissue before collagenase treatment (*see* **Note 4**).

3. Place tissues into collagenase solution for 30 min at 37 °C. After incubation, pipette tissue up and down gently for approximately 10 times to obtain a single-cell suspension (*see* **Note 2**).
4. Filter the suspensions using a 40-μm nylon cell strainer.
5. Add PBS to the suspension and centrifuge at 900 × *g* for 5 min.
6. For hemolysis, add lysing buffer to samples and wait for 15 min.
7. Centrifuge cells at 900 × *g* for 5 min and rinse with PBS. Resuspend the cell pellet in 100–200 μl PBS.
8. Add all color-conjugated antibodies to the single-cell suspension. Add 0.3 μl of each antibody per 1.0 × 10^6 cells.
9. Incubate samples on ice for at least 30 min (*see* **Note 3**).
10. Add PI buffer to the suspension to remove death cells by flow cytometer.
11. Endothelial cells are defined as CD31+/CD34+/Kit−/Ter119−/CD45− and mesenchymal cells CD31−/CD34−/Kit−/Ter119−/CD45−. Set the gate on flow cytometer for each population and begin isolation.
12. Collect isolated cells in RNA later for RNA extraction.

C. Fetal liver

Carry out all the procedures at room temperature unless otherwise specified.

1. Harvest embryos and dissect the fetal liver under binocular microscope to make sure no contamination from other organ is present (*see* **Note 1**).
2. Cut livers into small pieces using a scalpel (Fig. 1) (*see* **Note 4**).
3. Digest the livers with collagenase solution for 10–15 min at 37 °C (*see* **Note 2**). Use approximately 5 ml of collagenase for a litter (6–9) of day 13.5 embryos. Give a strong shake to the suspension every 5 min.
4. Control visually that all tissue is digested. Add a large volume of PBS-2% serum (volume of PBS approx. 5 times the volume of collagenase) and then filter through cell strainer.
5. Spin down the cells at 400 × *g* for 7 min and count cells.
6. Add all color-conjugated antibodies to the single-cell suspension at 4 °C for 30–45 min (*see* **Note 3**).

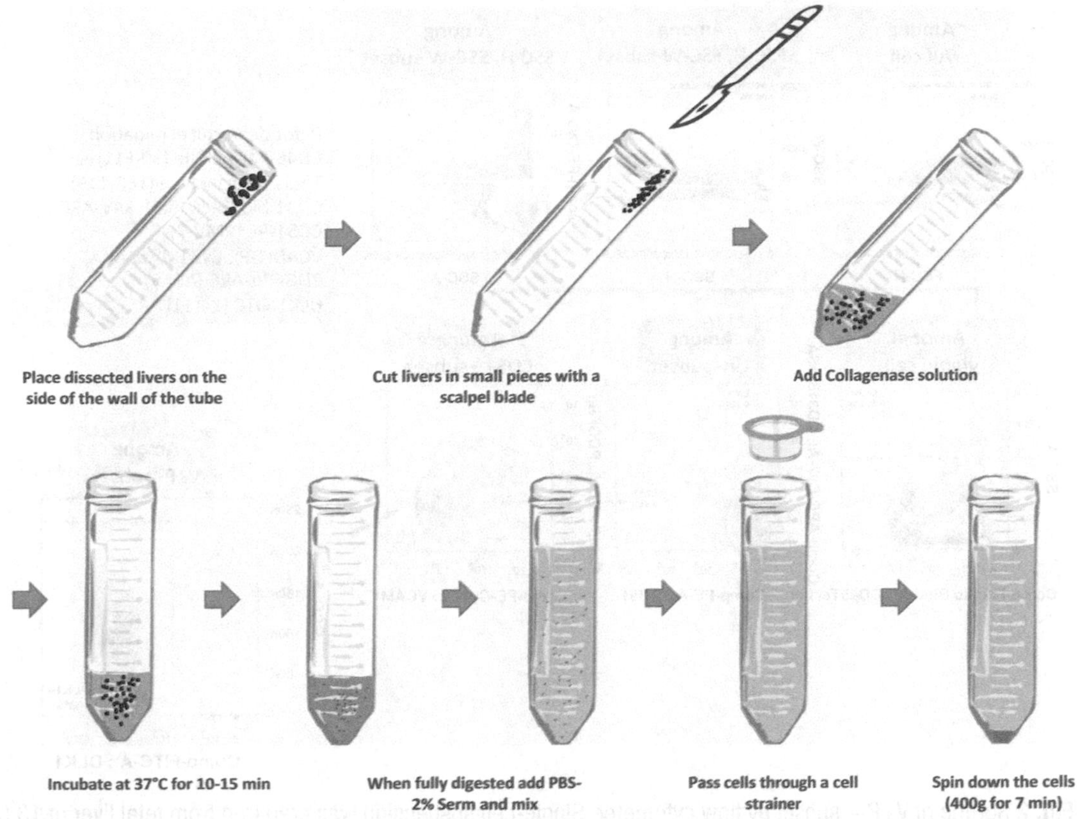

Fig. 1 Methodology for fetal liver cell preparation. Livers were placed on the side of the wall of the tube and cut into small pieces with a scalpel blade. A collagenase solution was added and incubated at 37 °C for 10–15 min. When the tissue has been fully digested, add PBS-2% serum and mix well. The cells were passed through a cell strainer and spun down at 400 × *g* for 7 min

7. Add PBS-2% serum to the suspension and centrifuge at 400 × *g* for 5 min twice. Incubated all tubes with biotinylated antibody with Step-PerCPCy5.5 (1/500).
8. Add PBS-2% serum to the suspension and centrifuge at 400 × *g* for 5 min twice.
9. Add PBS-2% serum-DAPI to remove death cells by flow cytometer.
10. Fetal liver stromal cells are defined as Ter119−/CD45−/CD31−/CD51+/VCAM-1+/PDGFRα- (V+P−), Ter119−/CD45−/CD31−/CD51+/VCAM-1−/PDGFRα+ (V−P+), and Ter119−/CD45−/CD31−/CD51+/VCAM-1+/PDGFRα+ (V+P+). Set the gate on flow cytometer for each population and begin isolation (Fig. 2).
11. Collect isolated cells in RNA later for RNA extraction.

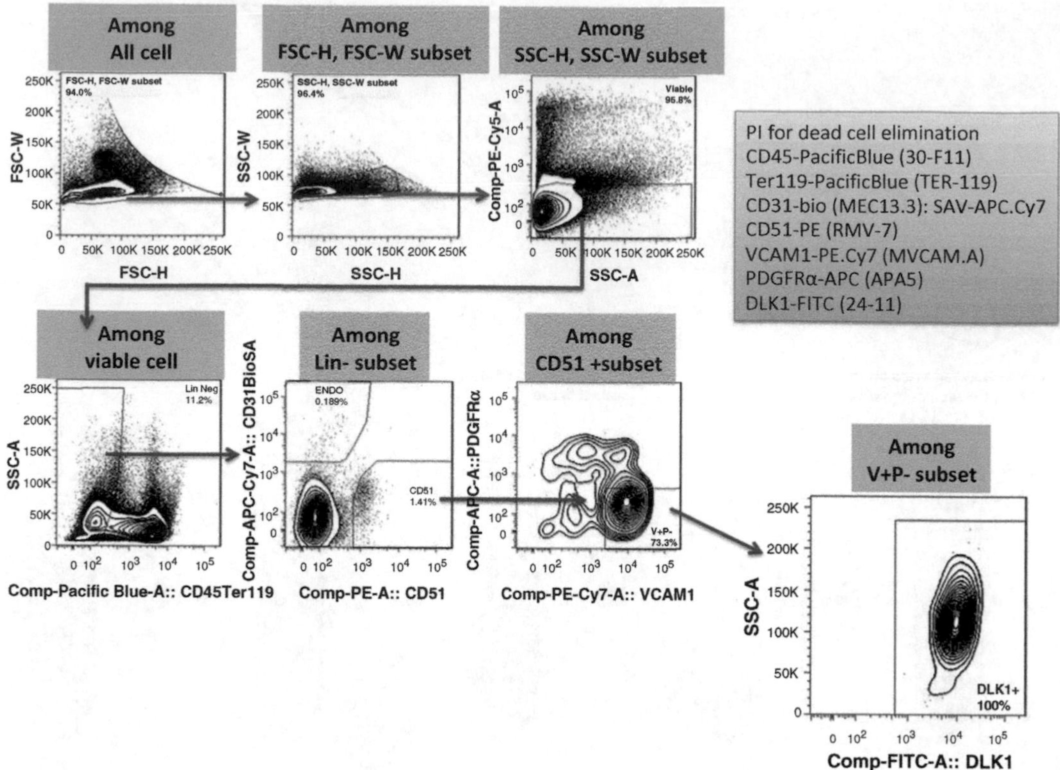

Fig. 2 Sorting of V+P− subset by flow cytometer. Single-cell suspension was prepared from fetal liver at 13.5 dpc. Fetal liver stromal cells are defined as Ter119−/CD45−/CD31−/CD51+/VCAM-1+/PDGFRα− (V+P−), Ter119−/CD45−/CD31−/CD51+/VCAM-1−/PDGFRα+ (V−P+), and Ter119−/CD45−/CD31−/CD51+/VCAM-1+/PDGFRα+ (V + P+). Among the V+P− subset, most of the cells are expressing DLK1, a hepatoblast-like marker

D. Fetal muscle

Carry out all the procedures at room temperature unless otherwise specified.

1. Obtain left and right femurs and muscle tissues surrounding those structures of 14.5 to 19.5 dpc mouse embryos. Trim the tissues from femurs (*see* **Note 1**).
2. Place tissues into collagenase solution for 20 min at 37 °C. After incubation, pipette tissue up and down gently for approximately 10 times to obtain a single-cell suspension (*see* **Note 2**).
3. Filter the suspensions using a 70-μm nylon cell strainer.
4. Add PBS to the suspension and centrifuge at 900 × *g* for 5 min.
5. Centrifuge cells at 900 × *g* for 5 min and rinse with PBS. Resuspend the cell pellet in 100–200 μl PBS.

6. Treat the cells with anti-mouse CD16/32 Fc binding blocker to avoid nonspecific antibody binding, for 10 min at 4 °C.
7. Add PBS to the suspension and centrifuge at 900 × *g* for 5 min.
8. Add all color-conjugated antibodies to the single-cell suspension. Add 0.3 μl of each antibody per 1.0 × 10^6 cells.
9. Incubate samples on ice for at least 30 min (*see* **Note 3**).
10. Add PI buffer to the suspension to remove death cells by flow cytometer.
11. Muscle hematopoietic cells are defined as CD45+/c-Kit+/Sca-1+/−. Set the gate on flow cytometer for each population and begin isolation.
12. Collect isolated cells in RNA later for RNA extraction or for May-Grünwald Giemsa staining.

E. Fetal spleen

Carry out all the procedures at room temperature unless otherwise specified.

1. Obtain fetal spleen from pregnant mother. Remove stomach from the spleen (*see* **Note 1**).
2. Place spleen tissues into collagenase solution for 20–30 min at 37 °C. After incubation, pipette tissue up and down gently for approximately 10 times to obtain a single-cell suspension (*see* **Note 2**).
3. Filter the suspensions using a 70-μm nylon cell strainer.
4. Add PBS to the suspension and centrifuge at 900 × *g* for 5 min.
5. Resuspend the cell pellet in 100–200 μl PBS.
6. Add all color-conjugated antibodies to the single-cell suspension. Add 0.3 μl of each antibody per 1.0 × 10^6 cells.
7. Incubate samples on ice for at least 30 min (*see* **Note 3**).
8. Add PI buffer to the suspension to remove death cells by flow cytometer.
9. Endothelial cells are defined as CD31+/CD34+/Kit−/Ter119−/CD45− and mesenchymal cells CD31−/CD34−/Kit−/Ter119−/CD45−/CD51+/−. Set the gate on flow cytometry for each population and begin isolation.
10. Collect isolated cells in RNA later for RNA extraction.

3.2 Immuno histochemistry

(A) p-Sp/AGM region, (B) placenta, (C) fetal Liver, (D) fetal muscle, and (E) fetal spleen

Carry out all the procedures at room temperature unless otherwise specified.

1. Dissect out the mouse embryos or the abovementioned sections of a mouse embryo (*see* **Note 1**) and fix in 2% paraformaldehyde in PBS overnight.
2. Equilibrate the spleen in 30% sucrose in PBS overnight.
3. Wash them in PBS.
4. Embed tissues in OCT compound and freeze in liquid nitrogen.
5. Slice tissues at 5–8 or 10 μm thickness using a cryostat.
6. Transfer to glass slides and dry thoroughly (*see* **Note 5**).
7. Wash the sections three times in PBS-0.05% Tween-10 (PBS-T) or PBS with 10 min each (*see* **Note 6**).
8. Block with blocking buffer for 30–45 min.
9. Incubate sections with appropriate dilutions of the primary antibodies at 4 °C overnight (*see* **Note 7**).
10. Wash samples in PBS-T or PBS three times for 30 min each (*see* **Note 6**).
11. Incubate sections with appropriate dilutions of the secondary antibodies overnight at 4 °C in a humidified chamber or at room temperature for 30 min (*see* **Note 8**).
12. Carry out all the procedure in darkroom after incubated with secondary antibody solution.
13. Wash samples in PBS-T or PBS three times for 30 min each.
14. Mount the samples on slides using fluorescent mounting medium, cover with cover glass.
15. Wait until the medium dries thoroughly and assess using a confocal microscope (*see* **Note 9**).

4 Notes

1. Ensure to remove excess tissues surrounding the hematopoietic organs to avoid contamination from the adjacent organs.
2. Collagenase shall be stocked at −30 °C and thawed before use.
3. Cells may be damaged and an adequate cell numbers may not be obtained during flow cytometric analysis, if the treatment time is exceeding 30 min.
4. When preparing samples containing large tissue fragments, an 18- rather than a 21-gauge needle is preferable. Alternatively, these tissues can be cut into smaller pieces using a razor/scalpel blade prior passage through the needle.

5. The tissue slice shall be dried overnight at room temperature to minimize the chance of detachment from the slide.
6. Gently apply any solutions onto the slides to avoid the tissue detached from the slide as those tissues are fragile and without undergoing fixation process.
7. Antibodies that are used for immunohistochemistry shall not be raised in the same host species to avoid cross-reactivity.
8. Avoid incubating longer than 30 min as it will cause high and nonspecific fluorescence signal during confocal microscopic analysis.
9. Slides can retain its fluorescence signal for approximately 1 month when kept at 4 °C in dark.

Acknowledgments

We thank the Ministry of Education, Culture, Sports, Science and Technology; The Ministry of Health, Labor and Welfare; and the Japan Society for the Promotion of Science for grant support.

References

1. Sugiyama D, Inoue-Yokoo T, Fraser ST, Kulkeaw K, Mizuochi C, Horio Y (2011) Embryonic regulation of the mouse hematopoietic niche. ScientificWorldJournal 11:1770–1780
2. Swain A, Inoue T, Tan KS, Nakanishi Y, Sugiyama D (2014) Intrinsic and extrinsic regulation of mammalian hematopoiesis in the fetal liver. Histol Histopathol 29:1077–1082
3. Suire C, Brouard N, Hirschi K, Simmons PJ (2012) Isolation of the stromal-vascular fraction of mouse bone marrow markedly enhances the yield of clonogenic stromal progenitors. Blood 119:e86–e95
4. Sasaki T, Mizuochi C, Horio Y, Nakao K, Akashi K, Sugiyama D (2010) Regulation of hematopoietic cell clusters in the placental niche through SCF/Kit signaling in embryonic mouse. Development 137:3941–3952
5. Sasaki T, Tanaka Y, Kulkeaw K, Yumine-Takai-A, Tan KS, Nishinakamura R, Ishida J, Fukamizu A, Sugiyama D (2016) Embryonic intra-aortic clusters undergo myeloid differentiation mediated by mesonephros-derived CSF1 in mouse. Stem Cell Rev 12:530–542
6. Sugiyama D, Kulkeaw K, Mizuochi C, Horio Y, Okayama S (2011) Hepatoblasts comprise a niche for fetal liver erythropoiesis through cytokine production. Biochem Biophys Res Commun 410:301–306
7. Sugiyama D, Kulkeaw K, Mizuochi C (2013) TGF-beta-1 up-regulates extra-cellular matrix production in mouse hepatoblasts. Mech Dev 130:195–206
8. Brouard N, Jost C, Matthias N, Albrecht C, Egard S, Gandhi P, Strassel C, Inoue T, Sugiyama D, Simmons PJ et al (2017) A unique microenvironment in the developing liver supports the expansion of megakaryocyte progenitors. Blood Adv 1:1854–1866
9. Tan KS, Kulkeaw K, Nakanishi Y, Sugiyama D (2017) Expression of cytokine and extracellular matrix mRNAs in fetal hepatic stellate cells. Genes Cells 22:836–844
10. Tan KS, Inoue T, Kulkeaw K, Tanaka Y, Lai MI, Sugiyama D (2015) Localized SCF and IGF-1 secretion enhances erythropoiesis in the spleen of murine embryos. Biol Open 4:596–607
11. Tanaka Y, Inoue-Yokoo T, Kulkeaw K, Yanagi-Mizuochi C, Shirasawa S, Nakanishi Y, Sugiyama D (2015) Embryonic hematopoietic progenitor cells reside in muscle before bone marrow hematopoiesis. PLoS One 10: e0138621

Methods in Molecular Biology (2019) 2002: 29–38
DOI 10.1007/7651_2018_175

Published online: 11 August 2018

Functional Assays of Stem Cell Properties Derived from Different Niches

Beatriz de Lucas, Laura M. Pérez, and Beatriz G. Gálvez

Abstract

It has been described that adult tissues contain mesenchymal stem cell populations. The specific areas where stem cells reside are known as niches. Crosstalk between cells and their niche is essential to maintain the correct functionality of stem cell. MSCs present a set of abilities such as migration, invasion, and angiogenic potentials, which make them ideal candidates for cell-based therapies. In order to test the regenerative capacity of these cells, we have described a methodology for the collection and for the evaluation of these mesenchymal precursors from different niches.

Keywords Angiogenesis, Explant, Invasion, Mesenchymal stem cell, Migration, Niche

1 Introduction

Regenerative properties of stem cells such as migration, invasion or angiogenesis are essential abilities for regenerative cell-based therapies. Mesenchymal stem cells (MSCs) seem to be good candidates for cell therapies since they have self-renewal capacity and present great potential to differentiate into multiple cell types. Multiple adult tissues contain this population of mesenchymal stem/progenitor cells [1–3], and they have been well characterized as progenitors residing in specific locations or niches into the adult tissues. Adult stem cells reside in stem cell niches that are necessary to maintain tissue homeostasis and repair by controlling stem cell behavior. It helps stem cell to carry on self-renewing divisions and it maintains stem cell's properties to be able to carry out its repair functions into the tissues.

It has been reported the presence of MSCs in almost every adult tissue allowing the possibility to obtain diverse stem cells from different sources [4]. Here, we provide a method to isolate MSCs from diverse adult tissue niches by the explant technique [1, 3]. The question of which stem cell source will be optimal for

Beatriz de Lucas and Laura M. Pérez contributed equally to this work.

each disease condition remains to be answered; hence, more research on the field is necessary.

We also propose a set of in vitro techniques to assess the regenerative ability of cells for their use on cell therapies. Migration is a key property in vivo essential to reach the area of interest into the tissues [5]. Evaluation of cell migration has been developed through a range of techniques and assays that allow to measure the capability of cells to move. Transwell system evaluates migration by using membranes with pores and wound-healing assay is used to assess cell motility in two dimensions. Additionally, it is necessary to assess if cells are able to cross a matrix and invade the tissue [6], so we described a method to evaluate the ability of cells to transmigrate by using the previous Transwell system covered by a 3D matrix. Finally, it has been shown that MSCs can participate in regeneration by helping in angiogenic processes [7]. Therefore, we enclose an easy method to measure angiogenic potential in cells. In this chapter, we aim to show a set of methods for isolation and functional characterization of cells residing on different niches.

2 Materials

Prepare and store all reagents and samples at room temperature (unless indicated otherwise). Use and prepare all materials in sterile conditions.

2.1 Mouse Samples

1. Adult C57BL/6 mice
2. Dissection tools: sterile surgical blade and handle, forceps, and scissors
3. 15-ml conical tubes placed on ice

2.2 Preparation of Explant Technique

1. Adult tissues collected from mouse
2. Prepared plates: Add 50 μl Matrigel (BD Biosciences, USA) around the well perimeter of 24-well plate (*see* **Note 1**). Incubate at 37 °C for 30 min
3. Culture microscope

2.3 Cell Culture

1. 24-Well plate and culture dishes
2. Complete medium consisting of Dulbecco's modified Eagle's medium (DMEM) (Sigma, USA) supplemented with 10% fetal bovine serum (FBS) (Sigma, USA), 105 U/mL penicillin/streptomycin (Lonza, Switzerland), 2 mM L-glutamine (Lonza, Switzerland), and 10 mM HEPES (Lonza, Switzerland)
3. Store the complete sterilized medium at 4 °C
4. Warm the medium to 37 °C before using it for culturing

5. Solution of trypsin–EDTA (0.25% trypsin and 0.02% EDTA) (Sigma, USA)

2.4 Migration Assay

1. 24-Well plate and 8-μm Transwell filters (Corning Incorporated, USA)
2. Fixative solution: 4% glutaraldehyde (Sigma, USA) in PBS
3. Staining solution: 2% toluidine blue (Sigma, USA) in water
4. Cotton swab and scalpel
5. Glycerol gelatin (Sigma, USA) mounting medium

2.5 Invasion Assay

1. 24-Well plate and 8-μm Transwell filters (Corning Incorporated, USA)
2. Extracellular matrix: Add 600 μl of 1% gelatin in the lower chamber, place on 24-well plate, and add 100 μl in the upper chamber. Allow to solidify in humidified incubator at 37 °C for 1 h to form a thin gel layer (*see* **Note 2**)
3. Fixative solution: 4% glutaraldehyde (Sigma, USA) in PBS
4. Staining solution: 2% toluidine blue (Sigma, USA) in water
5. Cotton swab and scalpel
6. Glycerol gelatin (Sigma, USA) mounting medium

2.6 Angiogenic Assays

1. Dilute Matrigel in cold medium (DMEM) and add 60 μl into a 96-well plate (*see* **Note 3**).
2. Incubate for 30 min at 37 °C to allow it to gel.

3 Methods

Adult mesenchymal precursors can be efficiently obtained from a great variety of tissues. In this chapter, we are going to focus only on the isolation of mesenchymal precursors from adipose tissue (adipose precursor, AP), cardiac tissue (cardiac precursor, CP), skin (skin precursor, SP), lung (lung precursor, LP), and skeletal muscle (skeletal muscle precursor, SMP). These precursors were used to perform angiogenesis and the different migration and invasion assays.

3.1 Mesenchymal Precursor Isolation from Diverse Niches, the Explant Technique

1. Adult tissues were collected from C57BL6 mice and kept in DMEM w/o FBS with antibiotics.
2. Under sterile conditions, rinse each adult tissue and dissect it into 1–2 mm pieces with a scalpel.
3. Examine the tissue fragments under a microscope to select those containing small blood vessels.

4. The tissue explants are placed in the center of 24-well plate previously coated with Matrigel and expose blood vessels at the explant border (*see* Subheading 2.2).
5. Carefully add 200 μl of complete medium to each well (*see* **Note 4**).
6. Incubate in a cell culture incubator at 37 °C with 5% CO_2.
7. Check the explant daily to control the appearance of a population of small, rounded, and refractive cells floating (*see* **Note 5** and Fig. 1).
8. Collect floating cells with a pipette into a 24-well plate coated-free with complete medium to start cell expansion.

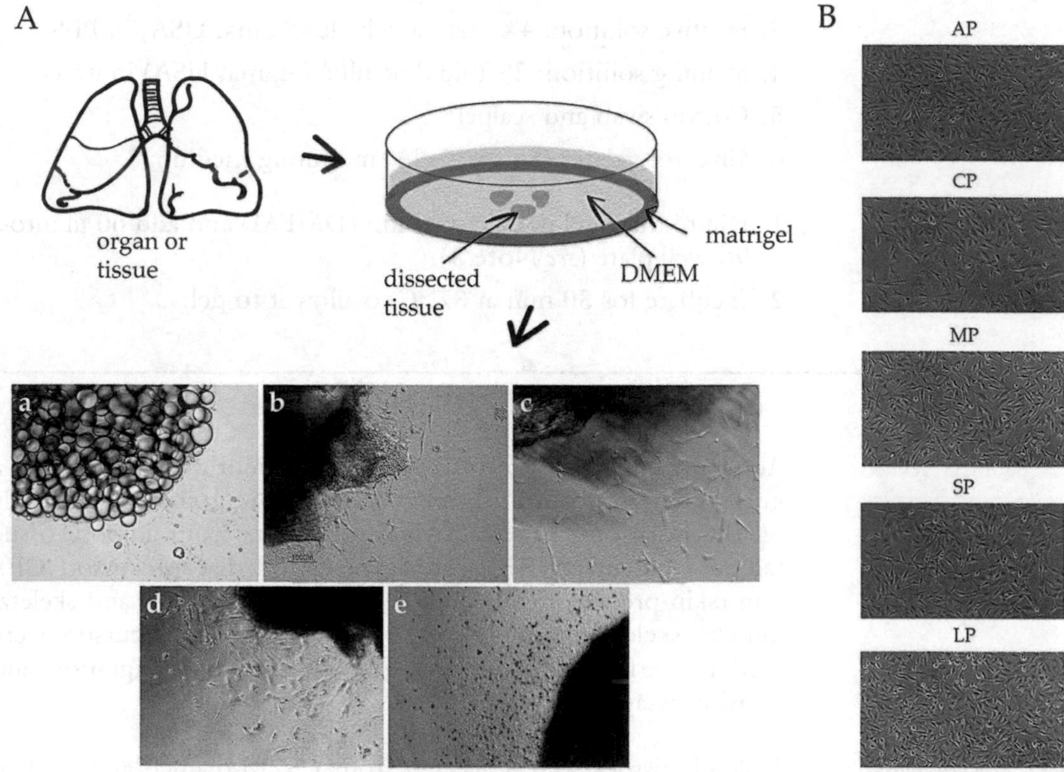

Fig. 1 Mesenchymal precursors isolated from adult tissues: adipose tissue, heart, skin, lung, and skeletal muscle. (**A**) Scheme of mesenchymal precursor isolation using the explant technique. Representative images of MSC emerging from explant in the different tissues: (**a**) subcutaneous white fat explant, (**b**) cardiac muscle explant with cells, (**c**) skeletal muscle explant with cells, (**d**) skin explant with cells, and (**e**) lung explant with cells. (**B**) Representative images of mesenchymal precursor cultures. *AP* adipose precursors, *CP* cardiac precursors, *MP* muscle precursors, *SP* skin precursors, *LP* lung precursors

3.2 Migration Measurement of Individual Cell or in Masses by In Vitro Scratch Assay (Wound-Healing Assay)

1. Culture mesenchymal precursors on a sterile 24-well plate with complete medium and place in a cell culture incubator maintained at 37 °C with 5% CO_2 to reach confluence (*see* **Note 6**).
2. Remove the medium and perform the wound with a sterile micropipette tip.
3. Wash with PBS to remove floating cells and photograph at time 0 h using the 4× magnification (*see* Fig. 2).
4. Add fresh complete medium and place in the cell culture incubator.

Sampling methodology: Take pictures at time 0 h and at different time points to follow migration. Use triplicates of each one.

Software: To measure cell migration through wound area measurement, use, for example, *ImageJ software*. By using this software, you can select the areas of interest and measure it.

Data processing: Migration rates are estimated as percentage of migration, the total area at a point time divided by the total area at time 0 h.

$$\text{Percentage of migration} = 100 - \left(\frac{\text{area at time } X \text{ h}}{\text{area at time 0 h}} \times 100\right)$$

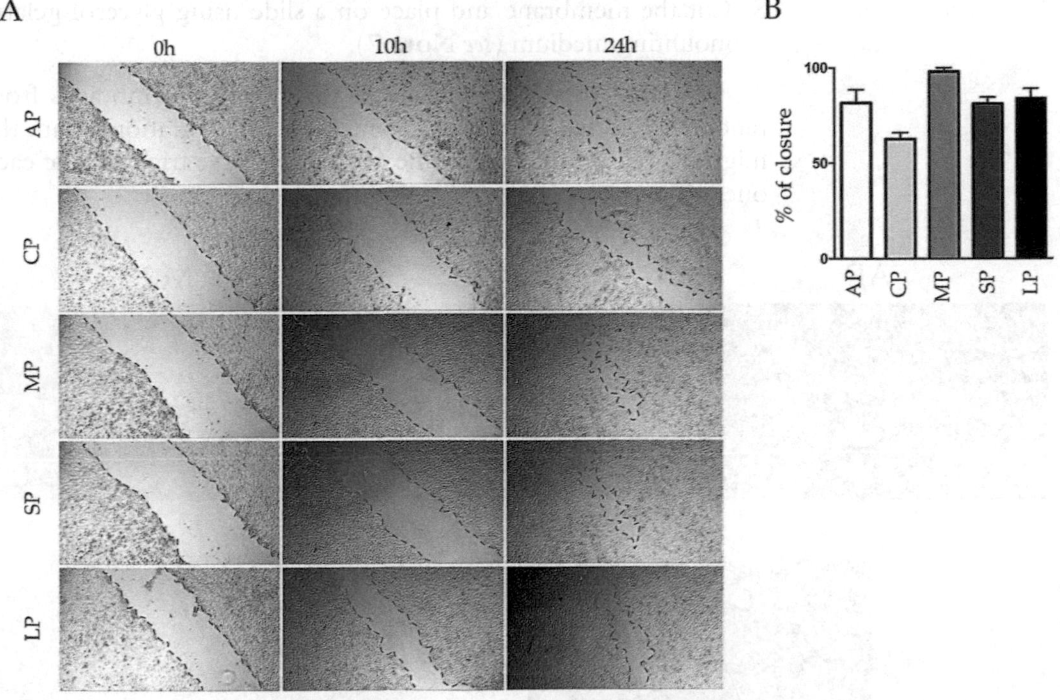

Fig. 2 Mesenchymal precursor migration by wound-healing assay. (**a**) Representative images of wound at time 0, 10, and 24 h from different precursors. (**b**) Representative graph of wound closure percentage from the different mesenchymal precursors. *AP* adipose precursors, *CP* cardiac precursors, *MP* muscle precursors, *SP* skin precursors, *LP* lung precursors

Statistical analysis: Data can be presented as the mean (±SEM) and analyzed with any statistical package software such as *GraphPad PRISM software* (GraphPad Software Inc., San Diego, CA). The analysis consists of one-way analysis of variance (ANOVA) with Bonferroni multiple comparison multiple group to determine the differences in cell migration between groups each time.

3.3 Cell Migration Assessment by Transwell Assay

1. Prepare a 24-well plate by adding 600 μl of complete medium.
2. Insert a Transwell into each well.
3. Seed cell suspension on the top of each Transwell, 2.5×10^4 cells in 80 μl of complete medium.
4. Place in a cell culture incubator maintained at 37 °C with 5% CO_2 for 24 h.
5. Fix the cells transferring the Transwell chamber to a new 24-well plate with 500 μl of 4% glutaraldehyde. Incubate at room temperature for 2 h.
6. Stain the cells transferring the Transwell chamber to a new 24-well plate with 500 μl of 2% toluidine blue. Incubate at room temperature overnight.
7. Wash Transwell chambers in water and clean the upper side with a cotton swab to eliminate nonmigratory cells.
8. Cut the membrane and place on a slide using glycerol-gelatin mounting medium (*see* **Note** 7).

Sampling methodology: Take pictures of the membranes from randomly selected 10× fields. To measure cell migration, count the migrated cells in the membrane (*see* Fig. 3). Use triplicates of each one.

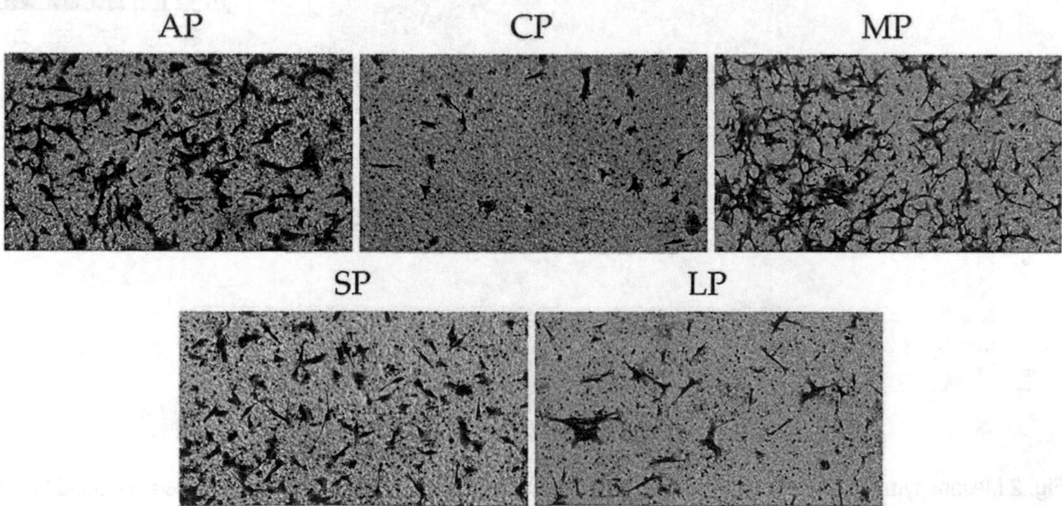

Fig. 3 Mesenchymal precursor migration by Transwell assay. Representative images of the migrated cells on a membrane of mesenchymal precursors. *AP* adipose precursors, *CP* cardiac precursors, *MP* muscle precursors, *SP* skin precursors, *LP* lung precursors

Data processing: Migration rates are estimated as migrated cells per field (*see* Fig. 3).

Statistical analysis: Data can be presented as the mean (±SEM) and analyzed with any statistical package software such as GraphPad PRISM software (GraphPad Software Inc., San Diego, CA). The analysis consists of one-way analysis of variance (ANOVA) with Bonferroni multiple comparison multiple group to determine differences between groups.

3.4 Cellular Invasion, the Ability of a Cell to Invade a 3-D Matrix (Fig. 4)

1. Use the previously prepared 24-well plate with gelatin.
2. Remove the gelatin from the lower and the upper chambers of the Transwell.
3. Add 600 μl of complete medium in the lower chamber and insert the Transwell.
4. Seed cell suspension on the top of each Transwell (2.5 × 10^4 cells in 80 μl of complete medium).
5. Place in a cell culture incubator maintained at 37 °C with 5% CO_2 for 24 h.
6. Fix, stain, and prepare the membranes as the previous Transwell migration assay (points 6–8).

Sampling methodology and statistical analysis: as in migration assay with Transwell

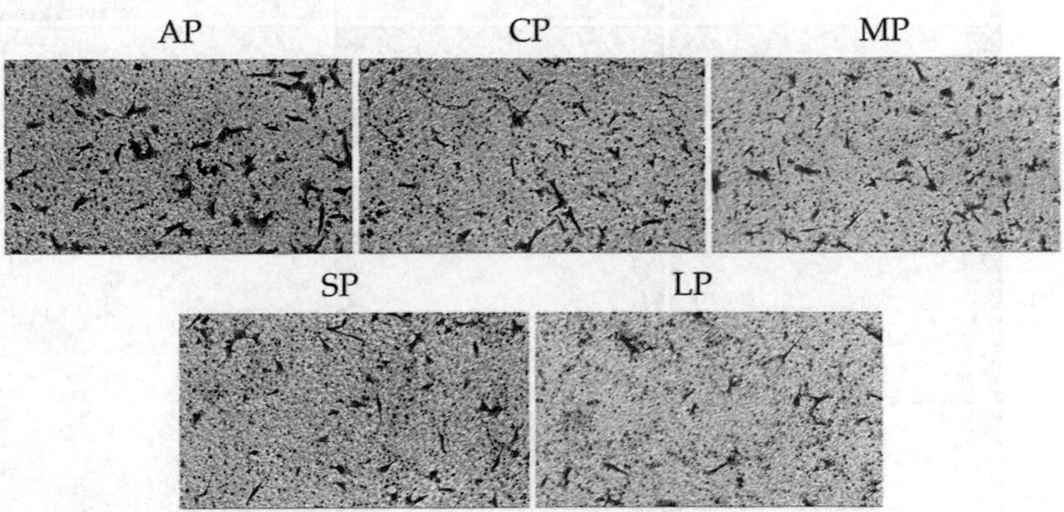

Fig. 4 Mesenchymal precursor invasion. Representative images of the invasive cells on a membrane of mesenchymal precursors. *AP* adipose precursors, *CP* cardiac precursors, *MP* muscle precursors, *SP* skin precursors, *LP* lung precursors

3.5 Tubular-Like Structures In Vitro, Angiogenesis Evaluation

1. Use the previously prepared 96-well plate with Matrigel.
2. Seed cell suspension on top of the Matrigel (5×10^3 cells in 60 μl of complete medium) (*see* **Note 8**).
3. Place in a cell culture incubator maintained at 37 °C with 5% CO_2.
4. Follow tubular-like structures formation at different time points and take pictures to select the best one.

Sampling methodology: Capillary tubules are quantified as branch point numbers (*see* Fig. 5). Use triplicates of each one.

Statistical analysis: Data can be presented as the mean (±SEM) and analyzed with any statistical package software such as GraphPad

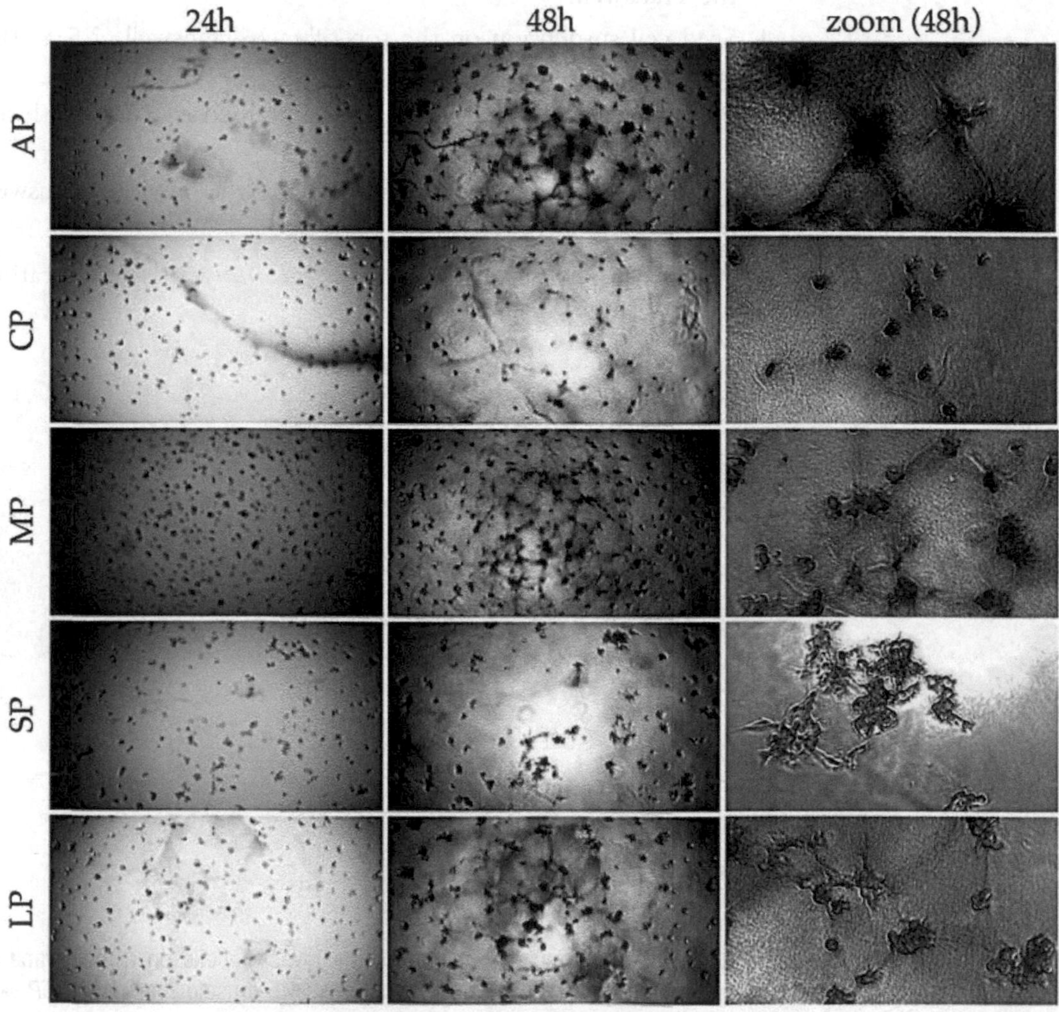

Fig. 5 Angiogenesis of mesenchymal precursors. Representative images of the capillary structures formed on Matrigel of mesenchymal precursors. *AP* adipose precursors, *CP* cardiac precursors, *MP* muscle precursors, *SP* skin precursors, *LP* lung precursors

PRISM software (GraphPad Software Inc., San Diego, CA). The analysis consists of one-way analysis of variance (ANOVA) with Bonferroni multiple comparison multiple group to determine differences between groups.

4 Notes

1. It is important to thaw Matrigel by submerging the vial in ice in a 4 °C refrigerator overnight prior to use. Keep it on ice at all times. Use cooled pipets, mix the Matrigel, and carefully place Matrigel around the well perimeter avoiding bubble formation.
2. After allowing to gel, remove carefully the gelatin excess from upper and lower chamber prior their use.
3. Add 1 part of Matrigel and 1 part of DMEM, mix carefully avoiding bubble, and place on a 96-well plate. Allow to gel for 30 min at 37 °C.
4. Complete medium must to be added carefully avoiding the possible movement of the tissue explant.
5. Check daily the explant, adding medium to keep it covered. In Fig. 1, some examples of different samples after several days in culture are shown. We can visualize the appearance of the cells emerging from the explants (*see* Fig. 1a). Appearance of the isolated cells after expansion (*see* Fig. 1b).
6. Cells can be seeded at a density between 2.0×10^4 and 1.0×10^4 cells (depending on the cell type) and incubated for 24 h in order to reach the confluence.
7. Put a drop of glycerol gelatin on a slide and pick up carefully the membrane placing the cells toward the slide.
8. Place the drop of cell suspension carefully in the center. Be careful not to touch the Matrigel.

Acknowledgments

This study was supported by the grant from the Spanish Ministry of Science and Innovation (SAF 2015–67911-R) to BGG. BdL is supported by FPU fellowships from the Spanish Ministry of Science and Innovation.

BdL and LMP wrote the manuscript. BGG revised the manuscript. All authors read and approved the final manuscript. We would like to thank all past and present members of the "*BGG group*", who have contributed to the design and evolution of these protocols over several years.

References

1. Bernal A, Fernandez M, Perez LM, San Martin N, Galvez BG (2012) Method for obtaining committed adult mesenchymal precursors from skin and lung tissue. PLoS One 7 (12):e53215. https://doi.org/10.1371/journal.pone.0053215
2. Galvez BG, Sampaolesi M, Barbuti A, Crespi A, Covarello D, Brunelli S, Dellavalle A, Crippa S, Balconi G, Cuccovillo I, Molla F, Staszewsky L, Latini R, Difrancesco D, Cossu G (2008) Cardiac mesoangioblasts are committed, self-renewable progenitors, associated with small vessels of juvenile mouse ventricle. Cell Death Differ 15(9):1417–1428. https://doi.org/10.1038/cdd.2008.75
3. Galvez BG, San Martin N, Rodriguez C (2009) TNF-alpha is required for the attraction of mesenchymal precursors to white adipose tissue in Ob/ob mice. PLoS One 4(2):e4444. https://doi.org/10.1371/journal.pone.0004444
4. Hass R, Kasper C, Bohm S, Jacobs R (2011) Different populations and sources of human mesenchymal stem cells (MSC): a comparison of adult and neonatal tissue-derived MSC. Cell Commun Signal 9:12. https://doi.org/10.1186/1478-811X-9-12
5. De Becker A, Riet IV (2016) Homing and migration of mesenchymal stromal cells: how to improve the efficacy of cell therapy? World J Stem Cells 8(3):73–87. https://doi.org/10.4252/wjsc.v8.i3.73
6. Weidenhamer NK, Moore DL, Lobo FL, Klair NT, Tranquillo RT (2015) Influence of culture conditions and extracellular matrix alignment on human mesenchymal stem cells invasion into decellularized engineered tissues. J Tissue Eng Regen Med 9(5):605–618. https://doi.org/10.1002/term.1974
7. Tao H, Han Z, Han ZC, Li Z (2016) Proangiogenic features of mesenchymal stem cells and their therapeutic applications. Stem Cells Int 2016:1314709. https://doi.org/10.1155/2016/1314709

Methods in Molecular Biology (2019) 2002: 39–50
DOI 10.1007/7651_2018_177

Published online: 04 September 2018

Ex Vivo Visualization and Analysis of the Muscle Stem Cell Niche

Aviva J. Goel and Robert S. Krauss

Abstract

Adult skeletal muscle stem cells, termed satellite cells, are essential for regenerating muscle after tissue damage. Satellite cells are located in a specialized microenvironment between muscle fibers and their surrounding basal lamina. This local niche serves as a compartment to preserve satellite cell function and provides signals that facilitate the rapid response to injury. Visualization of this local niche enables the elucidation of such niche-derived signals. Here, we describe techniques for isolating single myofibers with their associated satellite cells for ex vivo visualization and analysis of an intact muscle stem cell niche.

Keywords Cadherin, Catenin, Cell adhesion, Integrin, Muscle, Niche, Quiescence, Regeneration, Satellite cell, Stem cell

1 Introduction

Adult mammalian skeletal muscle has remarkable capacity to regenerate upon injury. This capability is enabled by resident muscle-specific stem cells, named satellite cells (SCs). SCs are generally quiescent during normal muscle homeostasis [1–3]. In response to tissue damage, SCs receive instructive cues that promote activation, rapid proliferation, migration, differentiation, and fusion to regenerate muscle tissue; they also self-renew [1–3]. Cues that promote quiescence vs. activation are provided in part by a specialized microenvironment where the SCs reside, called the SC niche. The SC niche is highly specialized, and its study has illuminated the idea that quiescence is not a default property of SCs, but rather a tightly regulated one, where signals from the niche help direct SC function and dynamics. Understanding the regulation of these mechanisms has immense implications in regenerative medicine, as emerging studies show that these processes are impaired in aging and in muscle diseases [4–9].

Within their immediate niche, SCs lay entrapped between the myofiber and the extracellular matrix (ECM), and display polarized apical and basal adhesive interactions. Apically, several cadherins (N-, M-, VE-cadherins, and probably others) are localized at the interface between the SC and the myofiber, forming classical cadherin/catenin-based adhesive junctions [3, 10, 11]. Basally, SCs adhere to the ECM largely through integrins, which bind laminins in the basal lamina [12, 13].

In vivo approaches used to identify and elucidate niche-derived signals have been challenging to employ, as the SC niche is difficult to visualize within a muscle cross-section. Additionally, niche-specific factors may be expressed at very low levels, making them difficult to visualize at the resolution with which one can examine a muscle section. These issues render such analyses cumbersome, time consuming, and expensive. An alternative approach used successfully by several groups involves ex vivo isolation of myofibers and their associated SCs [14–17]. In this approach, the extensor digitorum longus (EDL) muscle of mice is enzymatically and mechanically dissociated to isolate single myofibers with their associated SCs in a largely preserved myofiber niche. Here, we outline adapted techniques for: (1) myofiber isolation and culture that minimizes mechanical dissociation to better preserve the niche and SC dynamics, (2) fixation and staining of myofibers to better visualize the SC niche-based junctional components, and (3) analysis and quantification of the localization of components within the SC adhesive niche.

2 Materials

Prepare all cell culture reagents in a laminar flow hood using sterile technique. Follow all institutional waste disposal regulations for chemical and biological waste.

2.1 Myofiber Isolation Reagents

Myofibers are isolated and plated in low serum/growth-factor media to prevent aberrant activation of SCs during isolation.

1. Myofiber Plating Media (MPM): 1× Dulbecco's modified Eagle medium (DMEM; Gibco/Life Technologies/Thermo Fisher Scientific, Grand Island, NY, USA), with 10% heat-inactivated horse serum (HS, Life Technologies), and 1% 1M HEPES. Filter sterilize using a 0.22-μm sterile vacuum filtration system (EMD Millipore, Burlington, MA, USA) and store at 4 °C. This solution can be stored at 4 °C for up to 1 month
2. Type II Collagenase (Gibco/Life Technologies)
3. 100 mm cell culture dishes (NUNC/Thermo Fisher Scientific #151432). These dishes do not have a groove on the bottom and are ideally suited for visualizing individual myofibers on the dissection microscope
4. Round-bottomed Polystyrene 5 ml tubes (Falcon #08717006). These tubes have an etching at the 300 μl mark

5. Extra Fine 8.5 cm Bonn surgical scissors (Fine Science Tools, Foster City, CA, USA)
6. Dumont #5 Fine Straight Forceps (Fine Science Tools)
7. Standard Pattern Serrated Straight Forceps (Fine Science Tools)
8. Fire-polished Pasteur pipettes: Cut off the long, narrow end of a Pasteur pipette (Thermo Fisher Scientific) with a glass cutter or sharp blade. Light a Bunsen burner and slowly rotate the tip of the pipette in the flame until it becomes smooth. This ensures minimal damage to myofibers during trituration. Cool completely before use
9. Chick embryo extract, Ultrafiltrate (US Biological/Thermo Fisher Scientific)
10. Stereoscope or dissection microscope with light base
11. Temperature-controlled rotating water bath

2.2 Fixing and Staining Reagents

1. 1× Phosphate buffered saline (PBS)
2. Electron Microscopy Grade 16% paraformaldehyde (PFA, Electron Microscopy Sciences/Thermo Fisher Scientific #15710). Dilute to 8% in PBS, aliquot into 5 ml aliquots, and store at −20 °C until use. *See* **Note 1**
3. 0.2% Triton X-100 (Thermo Fisher Scientific) in PBS (PBTX)
4. Goat serum (GS, Gibco/Thermo Fisher Scientific)
5. Primary antibodies used: mouse IgG1 anti-Pax7-c (DSHB, 1:100), chick anti-syndecan 4 (gift from D. Cornelison, 1:100), mouse IgG1 anti-N-cadherin (D-4, Santa Cruz #sc-8424, 1:40), mouse IgG1 anti-M-cadherin (12G4, Santa Cruz #sc-81471, 1:50), mouse IgG1 anti-β-catenin (BD #610154, 1:500), rabbit anti-αT-catenin (Proteintech #13974-1-AP, 1:100), rabbit anti-γ-catenin (Cell Signaling #2309, 1:20), and mouse IgG1 anti-α7-integrin (3C12, MBL #K0046-3, 1:300). *See* **Note 2**
6. Secondary antibodies used: Goat anti-mouse IgG1 Alexa-647 (Life Technologies, 1:300), goat anti-chicken IgY Alexa-488 (Life Technologies, 1:300), goat anti-rabbit IgG Alexa-488 (Life Technologies, 1:300), goat anti-rabbit IgG Alexa-568 (Life Technologies, 1:300), and donkey anti-rabbit Alexa-647 IgG (Life Technologies, 1:300). *See* **Note 3**
7. Fluoroshield Mounting Medium with DAPI (Abcam, Cambridge, UK)
8. Superfrost Plus Precleaned Microscope Slides (Thermo Fisher Scientific)
9. 60 × 24 mm Glass coverslips (Thermo Fisher Scientific)

10. Hydrophobic barrier PAP pen (Vector Laboratories, Burlingame, CA, USA)
11. Optional: orbital platform shaker

2.3 Imaging and Analysis Components

1. Confocal microscope
2. ImageJ software (https://imagej.nih.gov/ij/download.html)

3 Methods

All procedures are carried out at room temperature unless otherwise indicated. Follow all institutional waste disposal regulations for disposing of waste materials.

3.1 Initial Preparation

1. Set the temperature of the water bath to 37 °C.
2. Make 2.4 mg/ml of Type II Collagenase in MPM. Use 800 μl per EDL muscle isolated. Aliquot into 1 ml microcentrifuge tubes and store in a water bath heated to 37 °C until use. This solution should be made fresh on the day of isolation.
3. Coat a 100-mm cell culture dish with HS by pouring in ~15 ml HS, swirling, and pouring the HS back into its receptacle. 1 × 100 mm dish is needed per EDL muscle isolated. Add 10 ml of MPM per plate in a dropwise fashion so not to disturb the HS coating. Store in a 37 °C incubator until use.

3.2 Isolation of Mouse EDL Muscle

The EDL muscle is situated at the front of the hindlimb, laying dorsolaterally from the tibialis anterior (TA) muscle. The ideal method for harvesting the EDL muscle is from tendon-to-tendon, such that individual myofibers remain intact and do not contract during enzymatic digestion. The EDL should only be handled by the tendons to avoid unnecessary manipulation of the muscle. The dissection should take no more than 5 min per EDL muscle.

1. Euthanize mouse according to institutional and IACUC guidelines.
2. Saturate the hindlimb and surrounding areas with 70% ethanol.
3. Secure the mouse on a dissection tray. We prefer securing the mouse on its side for the best angle of dissection (Fig. 1a, b).
4. Using serrated straight forceps, peel away the skin from the mouse hindlimb from the distal-to-proximal edge to expose the hindlimb muscles. Ensure that the skin is pulled below the ankle joint to reveal the necessary tendons. The TA muscle will be visible at this point (Fig. 1c, d).
5. Using fine straight forceps, peel away the fascia surrounding the TA muscle, from the knee joint to the ankle joint, taking

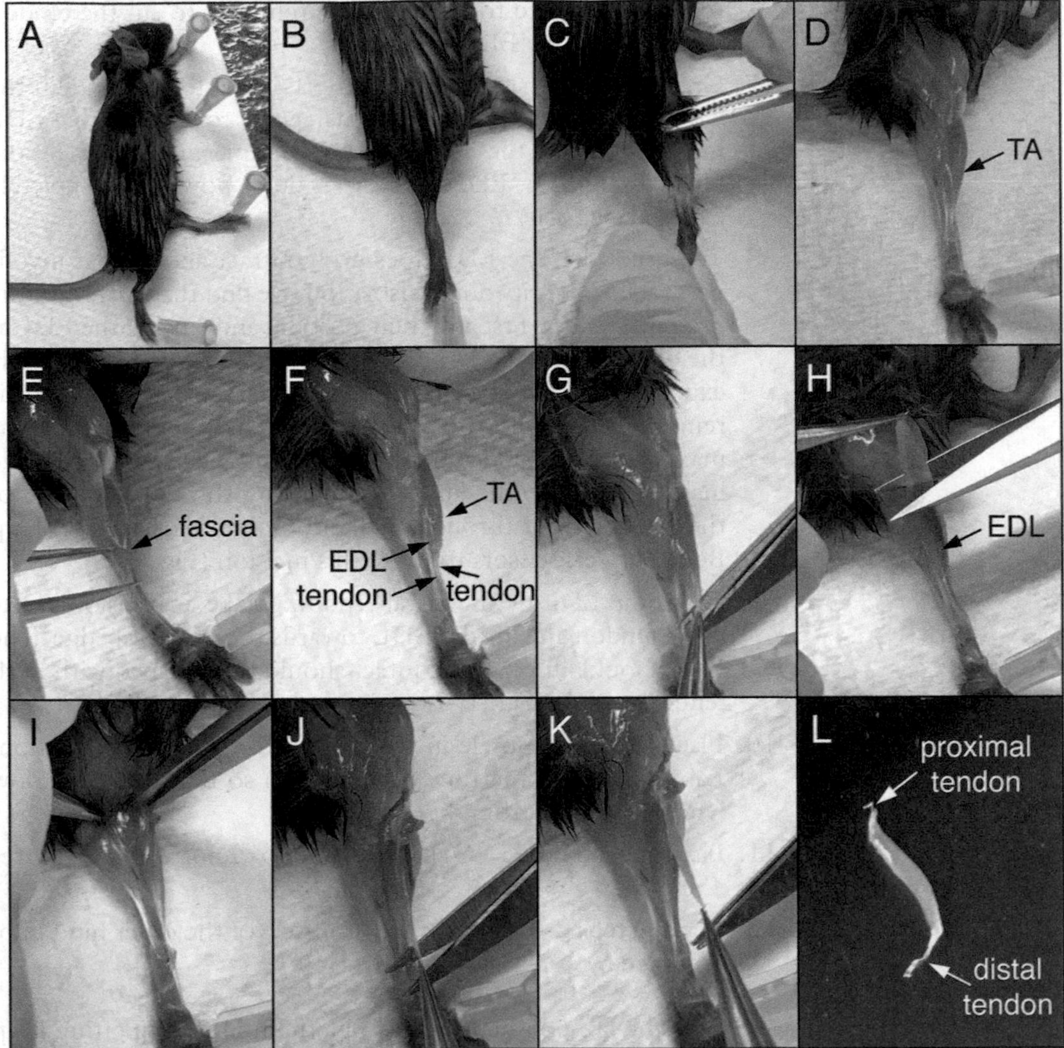

Fig. 1 Isolation of mouse EDL muscle. (**a**) Mouse secured on its side to the dissection tray. (**b**) Mouse hindlimb showing ideal angle for dissection. (**c**) Skin is peeled away from the hindlimb using serrated forceps. (**d**) Mouse hindlimb muscles. Arrow indicates the TA muscle. (**e**) Fascia surrounding the TA and EDL muscles is peeled away using fine forceps. (**f**) Mouse hindlimb muscles. Arrows indicate the TA, EDL, and their associated tendons. (**g**) The TA is grasped by its tendon and cut away. (**h**) Removal of the TA better reveals the EDL. (**i**) The proximal EDL tendon is cut by making an incision lateral to the knee joint. (**j**) The distal EDL tendon is grasped and cut. (**k**) The EDL is gently removed. (**l**) Intact isolation of the EDL showing proximal and distal tendons

care to not rupture the muscle itself. The fascia appears as a thin membrane surrounding the muscles (Fig. 1e).

6. At the distal edge by the ankle joint, two tendons are most prominent. The medial one connects to the TA muscle, and the lateral one to the EDL muscle. Using the fine forceps and Bonn surgical scissors, carefully make an incision at the medial

tendon connected to the TA. Be careful not to cut the tendon connected to the EDL (Fig. 1f, g).

7. Grasp the tendon connected to the TA with forceps and lift the TA muscle towards the proximal edge to expose the EDL muscle. Cut off the TA muscle at the proximal edge by the knee joint. The EDL muscle should now be completely exposed (Fig. 1h).
8. To ensure that the EDL does not contract and so that myofibers retain integrity during isolation, we find that cutting at the proximal tendon first and then grasping and cutting the EDL at the distal tendon works best. The proximal tendon is not easily exposed unless the connective tissue around the knee joint is removed. However, to maximize the speed of dissection (thus maximizing myofiber integrity), we find that making a large incision lateral to the knee joint, cutting through the edge of the bone but not through the entire joint, is ideal. Use sharp Bonn surgical scissors to make this incision (Fig. 1i).
9. Grasp the EDL by the distal tendon. Make an incision at the distal tendon and lift the EDL towards the proximal edge. The EDL should lift off and if intact should be S-shaped with both tendons visible (Fig. 1j–l).
10. Place the EDL muscle in the tube with collagenase solution. Label tube with the time of dissection so that you will know when to remove from digestion.
11. Place tube in the 37 °C shaking water bath for ~1 h. *See* **Notes 4** and **5**.
12. Repeat **steps 2–11** for the EDL muscle of the other hindlimb. *See* **Note 6**.

3.3 Isolation of Primary Myofibers from Mouse EDL Muscle

1. Coat the fire-polished Pasteur pipette in HS by pipetting it up and down.
2. After digestion, remove 1 × 100 mm dish containing MPM from the incubator and use HS-coated Pasteur pipette to transfer the EDL muscle from the tube into the dish.
3. *Critical step:* Viewing the muscle under a dissection microscope, use the Pasteur pipette to gently but firmly triturate the muscle against the wall of the plate for 5 min. Usually, 10–20 initial pulses are needed for individual myofibers to begin to dissociate from the muscle. There will also be a lot of muscle fragments that dissociate and contract. *See* **Notes 7** and **8**.
4. At this point, individual myofibers should be visible in the dish but can be wavy (Fig. 2a). Place dish back into the 37 °C incubator for 20 min to allow myofibers to straighten out (Fig. 2b). Myofibers with intact integrity will be long,

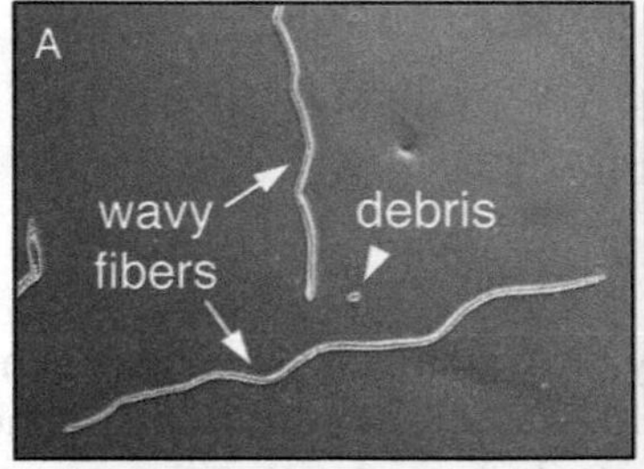

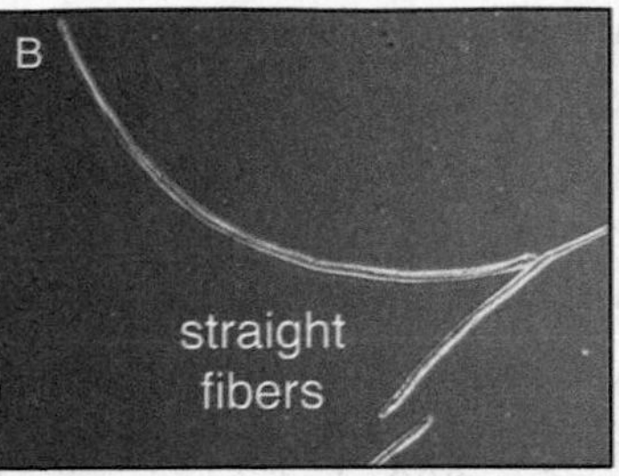

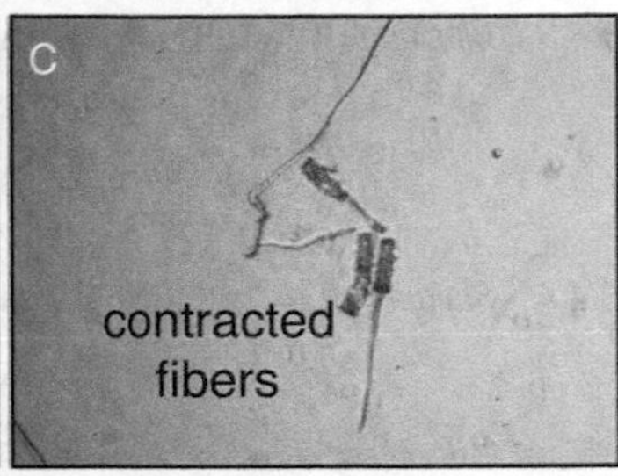

Fig. 2 Morphology of individual myofibers. (**a**) Individual myofibers immediately after isolation. Arrows indicate initial wavy morphology of fibers. Arrowhead indicates debris. (**b**) Myofibers upon incubation at 37 °C for 20 min. Note the straightened morphology of the fibers. (**c**) Contracted myofibers. Note the short, opaque, and kinked morphology

translucent, and refractile and have no visible kinks or tears. Contracted or partially contracted myofibers are opaque, short, and can have kinks (Fig. 2c).

5. Repeat **steps 2–4** for all isolated EDL muscles.
6. To culture myofibers, add 0.5% chick embryo extract to each 100 mm plate. We have cultured individual myofibers for up to 72 h to examine SC dynamics within their myofiber niche [10]. *See* **Note 9**.

3.4 Fixation and Staining of Myofibers

3.4.1 Fixation

1. Coat the round-bottomed Polystyrene 5 ml tubes with HS.
2. Using a P200 pipette with the tip coated in HS, transfer 20–25 single myofibers into each tube. Use caution to only transfer myofibers with intact integrity and without any kinks or breaks.
3. Wait for 5 min for fibers to settle to the bottom of the tube.
4. Aspirate the liquid only until the 300 μl mark on the tube so that no myofibers are aspirated off, and gently add 500 μl PBS to the tube to serially dilute out the MPM. Wait for 5 min, and aspirate again until the 300 μl mark on the tube.
5. Repeat PBS washes for 4 × 5 min for a total of five washes, aspirating again to the 300 μl mark on the tube.
6. Working in the chemical fume hood, add 300 μl of 8% PFA to each tube (therefore, the final concentration of PFA in each tube is 4%) and fix for 10 min in the dark (tent tubes with aluminum foil). *See* **Note 10**.
7. Wash fibers with 500 μl PBS for 4 × 5 min. At this stage, fixed fibers can be used immediately for staining or can be stored at 4 °C in the dark for up to 2 weeks.

3.4.2 Permeabilization

1. Aspirate liquid until the 300 μl mark on the tube.
2. Add 500 μl PBTX for 1 min.
3. Aspirate liquid until the 300 μl mark on the tube.
4. Add 500 μl PBTX for 10 min.

3.4.3 Blocking

1. Aspirate liquid until the 300 μl mark on the tube.
2. Add 10% GS to each tube (30 μl GS to each tube).
3. Block for 1 h at room temperature in the dark.

3.4.4 Staining

1. Aspirate liquid until the 300 μl mark on the tube.
2. Add primary antibodies directly into each tube at the concentration indicated (e.g., for 1:100, add 3 μl of antibody to the existing 300 μl of solution in the tube).
3. Incubate overnight at 4 °C (ideally on an orbital platform shaker).
4. Aspirate antibody solution to the 300 μl mark and wash 1 × 5 min with 500 μl PBS.
5. Wash 2 × 5 min with 500 μl PBTX.
6. Aspirate to the 300 μl mark and block for 30 min in 10% GS (i.e., add 30 μl GS to each tube).
7. Add secondary antibodies directly into each tube at the concentration indicated (e.g., for 1:300, add 1 μl of antibody to the existing 300 μl of solution in the tube).
8. Incubate with secondary antibody for 1 h at room temperature in the dark (ideally on an orbital platform shaker).
9. Aspirate secondary antibody solution to the 300 μl mark and wash 1 × 5 min with 500 μl PBS.
10. Wash 2 × 5 min with 500 μl PBTX.
11. Aspirate solution to the 300 μl mark.
12. Label a charged microscope slide for each tube.
13. Outline slides using a hydrophobic barrier PAP pen.
14. Take a P1000 pipet with the tip coated in HS, and transfer the contents of each tube onto its corresponding slide.
15. Allow myofibers to settle on the charged slides for 10–30 min in the dark.
16. Aspirate all liquid carefully, and wash slides 3 × 5 min with 200 μl PBTX, being careful not to aspirate any myofibers off the slides. *See* **Note 11**.
17. Aspirate all liquid carefully, and add two drops of DAPI-containing mounting medium to each slide. *See* **Note 12**.
18. Carefully cover slide with glass coverslip.
19. Allow slides to dry at room temperature for 2–4 h, seal coverslip with nail polish, and store at 4 °C.
20. Allow sealed slides to dry at least 24 h at 4 °C before imaging.

3.5 Imaging and Analyzing Junctional Components of the Adhesive Niche

Components of the adhesive niche are best imaged under a confocal microscope. *See* Fig. 3a, b for representative immunofluorescence images on EDL myofibers. We use a Leica SP5 DMI inverted quad laser scanning confocal microscope. Representative *z*-plane images should be exported to ImageJ for analysis of junctional localization. To analyze niche localization, we quantify the ratio of apical-to-basal signal intensity of junctional proteins across the SC niche. Single images across a *z*-plane are analyzed and the average intensity is computed. All images must be unprocessed for analyses.

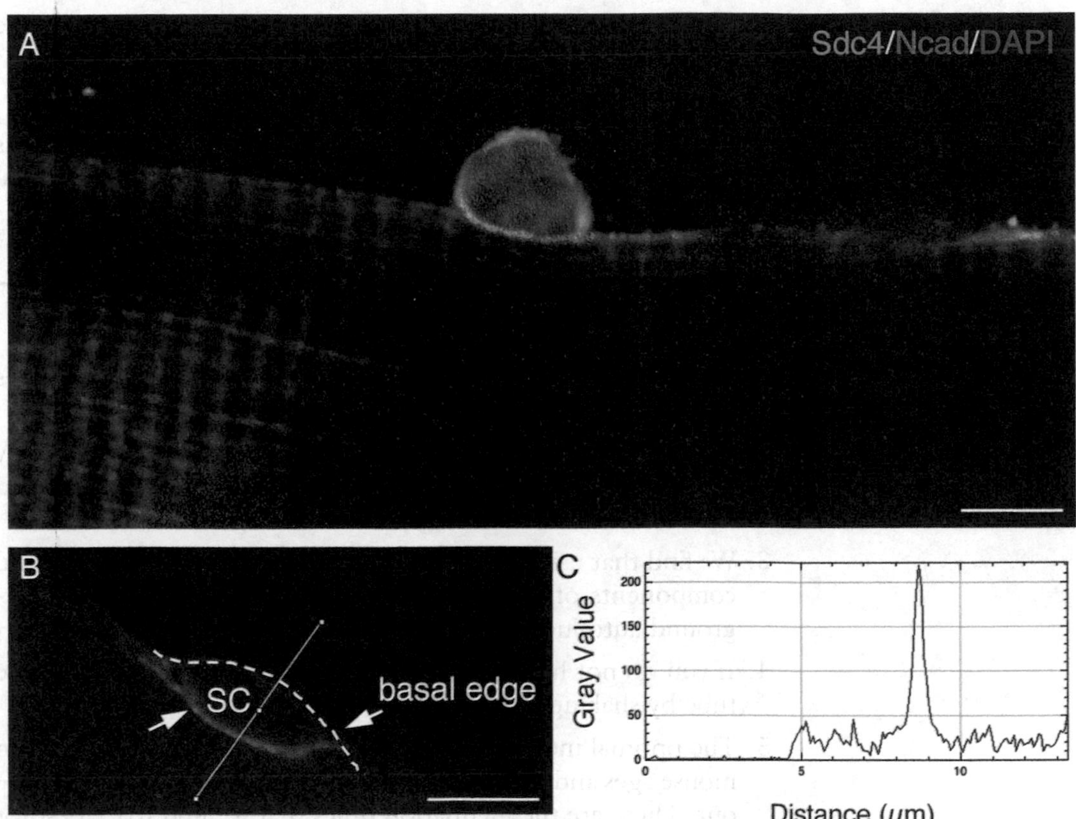

Fig. 3 Immunofluorescent visualization and quantification of a junctional component of the myofiber-SC niche. (**a**) Representative confocal immunofluorescence image of an EDL myofiber stained for N-cadherin (magenta), the SC marker, Sdc4 (green), and nuclei with DAPI (blue). Note that the background autofluorescence highlights sarcomeres in the myofiber (green stripes). (**b**) Representative confocal immunofluorescence image of β-catenin localization at the SC niche. Arrows indicate apical and basal edges. A straight line is drawn through the center of the SC for localization analysis. Note that the SCs on freshly isolated myofibers can adopt somewhat variable morphologies, ranging from more rounded (**a**) to more elongated (**b**). (**c**) Representative output for the plot profile of the line drawn in (**a**). Scale bars: 5 μm

1. Open the image of the individual channel for the protein to be measured in ImageJ.
2. Using the line tool, draw a straight line through the middle of the SC, across the apical and basal edges of the niche (Fig. 3b).
3. Compute the plot profile of the line. You can do this through the main menu by using your cursor to click on Analyze → Plot Profile.
4. The plot profile of the line will be computed as a Gray Value of the immunofluorescence signal intensity per unit distance (Fig. 3c).
5. Repeat for all images across the *z*-stack.
6. Repeat for at least 5–10 random SCs per mouse.
7. Export the Gray Value at the apical and basal edges of the niche to an Excel spreadsheet.
8. Compute the apical/basal ratio of intensity.
9. We plot these values on a $\log_2$ scale such that positive values indicate apical localization, and negative values indicate basal localization.

4 Notes

1. We find that using electron microscopy grade PFA minimizes background autofluorescence seen in myofibers.
2. Isotype-specific secondary antibodies were used for all primary antibodies with isotypes listed. This increased specificity and resulted in cleaner staining with less background.
3. We find that using 647-conjugated fluorophores for junctional components of the adhesive niche minimizes myofiber background autofluorescence.
4. If you do not have a shaking water bath, physically agitate the tube by shaking it every 5 min.
5. The optimal incubation time in the water bath varies between mouse ages and strains, and may require trial and error to figure out. These are the incubation times that we find work best for our mixed-strain mice:
   - Mice younger than 21 days—40 min
   - 1–2-month-old mice—50 min
   - 2–6-month-old mice—60 min
   - 6-month–1-year-old mice—70 min
   - Aged mice—90 min
   - Regenerated muscle—65–70 min

6. We recommend isolating fibers from no more than three mice (i.e., 6 EDL muscles) per preparation. This ensures that the time outside the incubator is kept to a minimum in-between individual muscle manipulation to avoid myofiber contraction.
7. If after 20 pulses there are no individual fibers dissociating from the EDL muscle, place the muscle back in the collagenase digestion tube and incubate for another 10 min and then try again. This can only be done once.
8. We find that 5 min of trituration is more than sufficient to dissociate myofibers with maximal integrity. Any additional mechanical manipulation may cause myofiber contraction and SC activation.
9. Other protocols suggest transferring individual myofibers to new culture plates with fresh media and added growth factors. We find that the additional manipulation can promote myofiber contraction. Instead, we simply add the growth factor-rich chick embryo extract directly to the original culture plate.
10. We find that fixing myofibers in the dark minimizes background autofluorescence.
11. We find that the easiest way to aspirate only the liquid and not the fibers is to place the aspirator on a corner of the slide and slowly aspirate off all liquid.
12. We find that it is best to mount each slide one-by-one to prevent myofibers from drying out.

Acknowledgments

We thank D. Cornelison, A. Brack, and C. Crist for gifts of reagents and helpful advice. This work was supported by NIH grants AR046207 and AR070231 to R.S.K. and by the Tisch Cancer Institute at Mount Sinai NIH P30 CA196521 for support of work on the Mount Sinai Microscopy CoRE.

References

1. Brack AS, Rando TA (2012) Tissue-specific stem cells: lessons from the skeletal muscle satellite cell. Cell Stem Cell 10(5):504–514. https://doi.org/10.1016/j.stem.2012.04.001
2. Dumont NA, Bentzinger CF, Sincennes MC, Rudnicki MA (2015) Satellite cells and skeletal muscle regeneration. Compr Physiol 5 (3):1027–1059. https://doi.org/10.1002/cphy.c140068
3. Dumont NA, Wang YX, Rudnicki MA (2015) Intrinsic and extrinsic mechanisms regulating satellite cell function. Development 142 (9):1572–1581. https://doi.org/10.1242/dev.114223
4. Shavlakadze T, McGeachie J, Grounds MD (2010) Delayed but excellent myogenic stem cell response of regenerating geriatric skeletal muscles in mice. Biogerontology 11 (3):363–376. https://doi.org/10.1007/s10522-009-9260-0
5. Chakkalakal JV, Jones KM, Basson MA, Brack AS (2012) The aged niche disrupts muscle stem cell quiescence. Nature 490

(7420):355–360. https://doi.org/10.1038/nature11438

6. Conboy IM, Conboy MJ, Smythe GM, Rando TA (2003) Notch-mediated restoration of regenerative potential to aged muscle. Science 302(5650):1575–1577. https://doi.org/10.1126/science.1087573
7. Joe AW, Yi L, Natarajan A, Le Grand F, So L, Wang J, Rudnicki MA, Rossi FM (2010) Muscle injury activates resident fibro/adipogenic progenitors that facilitate myogenesis. Nat Cell Biol 12(2):153–163. https://doi.org/10.1038/ncb2015
8. Dimmeler S, Ding S, Rando TA, Trounson A (2014) Translational strategies and challenges in regenerative medicine. Nat Med 20 (8):814–821. https://doi.org/10.1038/nm.3627
9. Lane SW, Williams DA, Watt FM (2014) Modulating the stem cell niche for tissue regeneration. Nat Biotechnol 32(8):795–803. https://doi.org/10.1038/nbt.2978
10. Goel AJ, Rieder MK, Arnold HH, Radice GL, Krauss RS (2017) Niche cadherins control the quiescence-to-activation transition in muscle stem cells. Cell Rep 21(8):2236–2250. https://doi.org/10.1016/j.celrep.2017.10.102
11. Krauss RS, Joseph GA, Goel AJ (2017) Keep your friends close: cell-cell contact and skeletal myogenesis. Cold Spring Harb Perspect Biol 9 (2). https://doi.org/10.1101/cshperspect.a029298
12. Rozo M, Li L, Fan CM (2016) Targeting beta1-integrin signaling enhances regeneration in aged and dystrophic muscle in mice. Nat Med 22(8):889–896. https://doi.org/10.1038/nm.4116
13. Kuang S, Gillespie MA, Rudnicki MA (2008) Niche regulation of muscle satellite cell self-renewal and differentiation. Cell Stem Cell 2 (1):22–31. https://doi.org/10.1016/j.stem.2007.12.012
14. Pasut A, Jones AE, Rudnicki MA (2013) Isolation and culture of individual myofibers and their satellite cells from adult skeletal muscle. J Vis Exp 73:e50074. https://doi.org/10.3791/50074
15. Vogler TO, Gadek KE, Cadwallader AB, Elston TL, Olwin BB (2016) Isolation, culture, functional assays, and immunofluorescence of myofiber-associated satellite cells. Methods Mol Biol 1460:141–162. https://doi.org/10.1007/978-1-4939-3810-0_11
16. Moyle LA, Zammit PS (2014) Isolation, culture and immunostaining of skeletal muscle fibres to study myogenic progression in satellite cells. Methods Mol Biol 1210:63–78. https://doi.org/10.1007/978-1-4939-1435-7_6
17. Kollu S, Abou-Khalil R, Shen C, Brack AS (2015) The spindle assembly checkpoint safeguards genomic integrity of skeletal muscle satellite cells. Stem Cell Reports 4 (6):1061–1074. https://doi.org/10.1016/j.stemcr.2015.04.006

Methods in Molecular Biology (2019) 2002: 51–59
DOI 10.1007/7651_2018_187

Published online: 30 August 2018

Stem Cell-Derived Cardiac Spheroids as 3D In Vitro Models of the Human Heart Microenvironment

Madeline Campbell, Mamta Chabria, Gemma A. Figtree, Liudmila Polonchuk, and Carmine Gentile

Abstract

Our laboratory has recently developed a novel three-dimensional in vitro model of the human heart, which we call the vascularized cardiac spheroid (VCS). These better recapitulate the human heart's cellular and extracellular microenvironment compared to the existing in vitro models. To achieve this, human-induced pluripotent stem cell (iPSC)-derived cardiomyocytes, cardiac fibroblasts, and human coronary artery endothelial cells are co-cultured in hanging drop culture in ratios similar to those found in the human heart in vivo. The resulting three-dimensional cellular organization, extracellular matrix, and microvascular network formation throughout the VCS has been shown to mimic the one present in the human heart tissue. Therefore, VCSs offer a promising platform to study cardiac physiology, disease, and pharmacology, as well as bioengineering constructs to regenerate heart tissue.

Keywords 3D cultures, Bioprinting, Cardiovascular regeneration, Heart microenvironment, Induced pluripotent stem cells, Niche, Tissue bioengineering, Vascularized cardiac spheroid

1 Introduction

Three-dimensional in vitro culture systems of the human heart cells have been developed to better recapitulate the heart's cellular and extracellular microenvironment or niche [1, 2]. Cells cultured in a 3D environment exhibit prolonged viability and improved physiological function when compared to cells cultured in 2D monolayer culture [3–6]. These differences are due to 3D culture systems more closely replicating the in vivo tissue niche, including cell–cell and cell–extracellular matrix (ECM) interactions, ECM composition and architecture, as well as gradients of oxygen, cytokines, and growth factors [1, 2, 6]. Our laboratory has developed vascularized cardiac spheroids (VCSs) by co-culturing human-induced pluripotent stem cell (iPSC)-derived cardiomyocytes, cardiac fibroblasts, and human coronary artery endothelial cells in hanging drop culture in ratios approximating those found in the human heart in vivo [7]. The nonadherent nature of hanging drop cultures promotes cell–cell adhesion and aggregation into the 3D spheroid culture

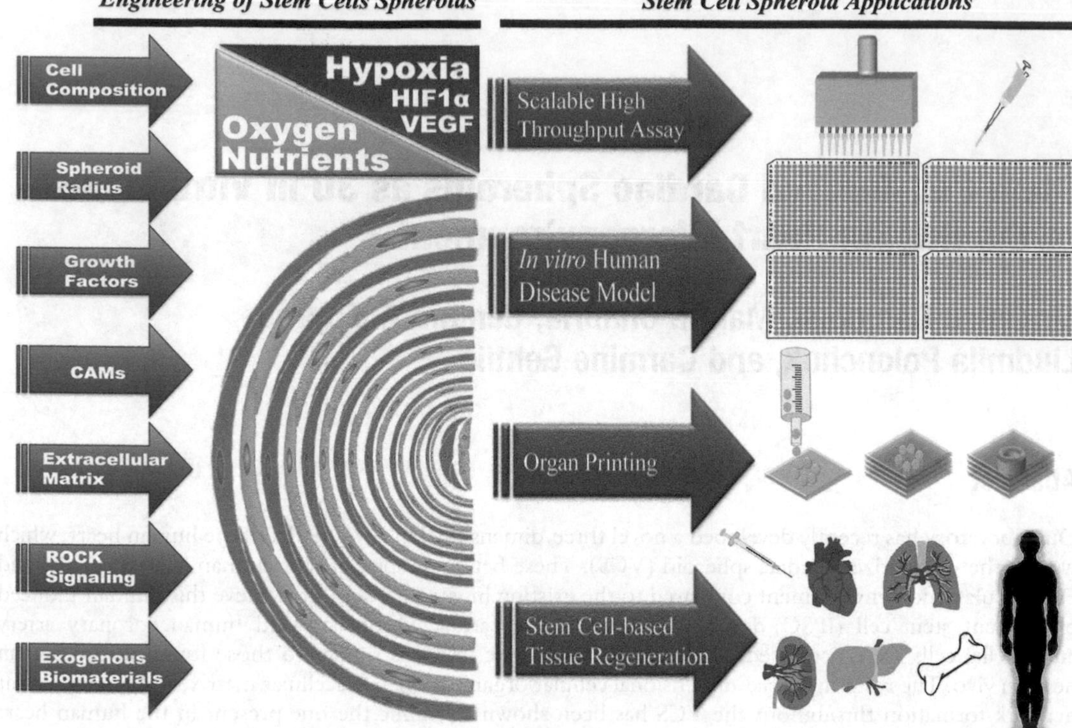

Fig. 1 Bioengineering the stem cell niche using spheroids for a range of in vitro and in vivo applications [1]

[1], in which the resulting cellular organization, extracellular matrix, and microvascular network formation has been shown to mimic the one present in the human heart [7]. Potential in vitro applications for VCSs include studying human heart development and physiology, modeling disease (e.g., cardiac fibrosis), and drug testing [7–10] (Fig. 1). Furthermore, VCSs could be used as building blocks for the engineering of more complex tissues in combination with 3D bioprinting technology for heart tissue regeneration in vivo [1, 11] (Fig. 1).

2 Materials

Prepare all solutions in biological safety cabinet under sterile conditions on ice (unless indicated otherwise).

2.1 Cell Culture

1. Human iPSC-derived cardiomyocytes (iCMs) (Ncardia, Cologne, Germany)
2. Human iPSC-derived cardiac fibroblasts (iCFs) (Ncardia, Cologne, Germany)

3. Human coronary artery endothelial cells (HCAECs) (Cell Applications, Inc., San Diego, CA, USA)
4. T25 cell culture flasks
5. Cor.4U Culture Medium (Ncardia)
6. 10 mg/mL Puromycin (Ncardia)
7. Fibronectin from bovine plasma (Sigma-Aldrich)
8. Human MesoEndo Cell Growth Medium (Cell Applications, Inc.)
9. Sterile 1× phosphate-buffered saline (PBS)
10. Trypsin–EDTA (0.25%) (Gibco)
11. Trypan blue (Gibco)
12. Falcon tubes
13. Centrifuge

2.2 CS Formation and Isolation

1. CS medium: prepared by combining Cor.4U Culture Medium (Ncardia) and MesoEndo Cell Growth Medium (Cell Applications, Inc.) (1:1 ratio)
2. 96-Well Hanging Drop Plate (3D Biomatrix, Ann Arbor, MI, USA)
3. Sterile stainless steel scissors
4. Sterile 1× PBS
5. Centrifuge
6. Falcon tubes

2.3 Fixation and Immunolabeling for Microscopy

1. Fixative: 1× PBS containing 4% paraformaldehyde
2. PBSA: 1× PBS containing 0.01% sodium azide
3. Permeabilization reagent: PBSA containing 0.02% Triton-X-100
4. Blocking solution: 3% bovine serum albumin (BSA) in PBSA
5. Primary antibodies: mouse monoclonal anti-human CD31/PECAM (BD Pharmingen, San Diego, CA, USA), mouse monoclonal [1C11] anti-human cardiac troponin T (Texas Red®) (Abcam, Cambridge, MA, USA), and mouse monoclonal [V9] to vimentin (Alexa Fluor® 488) (Abcam, Cambridge, MA, USA)
6. Secondary antibody: donkey anti-mouse secondary fluorochrome-conjugated antibodies (Jackson Immunological Research Labs, Inc., West Grove, PA, USA)
7. DNA stain: NucBlue® Live ReadyProbes® Reagent (Hoechst 33342) (Invitrogen, Carlsbad, CA, USA)

8. Glass-bottom plates or Petri dishes
9. Mounting medium

3 Methods

Carry out all procedures in biological safety cabinet under sterile conditions.

3.1 Cell Preparation

1. Precoat T25 flasks with 1 mL of PBS containing 40 μg/mL of fibronectin by incubating at 37 °C/5% CO_2 and removing the excess solution after 4 h.
2. Thaw iCMs, iCFs, and HCAEC cryovials in 37 °C water bath until only small ice crystals remain.
3. Plate each cell type in a separate T25 flask according to supplier's instructions. Culture iCMs and iCFs in Cor.4U Culture Medium on fibronectin-precoated T25 flasks. Culture HCAECs Human MesoEndo Cell Growth Medium in T25 flasks. Aliquot 3 mL medium into the T25 flask. Transfer contents of each cryovial into separate T25 flasks. To collect all cellular material, wash each cryovial with 1 mL aliquot of culture medium and repeat three times (*see* **Note 1**). The final volume in each T25 flask will be 7 mL in total. Allow HCAECs to adhere to culture surface overnight.
4. A final concentration of 5 μg/mL puromycin in Cor.4U Culture Medium is added to the T25 containing iCMs for cell selection. iCMs should survive this process.
5. After 24 h, aspirate the cell culture medium and wash once with 1× PBS to remove toxic cryopreservant. Aspirate PBS and immediately add 7 mL of appropriate culture medium to each T25 flask and incubate at 37 °C. Use puromycin-free Cor.4U Culture Medium for iCMs from this day onwards.
6. Check cell growth and viability daily and change culture medium every 1–2 days until cells reach confluency.
7. Once confluent, aspirate culture medium. Add 3 mL 1× PBS to wash cell culture by gently tilting the flask. Aspirate PBS and add trypsin–EDTA (0.25%) (3 mL for HCAECs and 1 mL for iCMs and iCFs) and incubate at 37 °C for 5 min for HCAECs and 3 min for iCMs and iCFs. Observe detached cells and gently tap the bottom of the flask to detach any residual cells. Immediately add 4 mL culture medium to each flask and transfer cell suspension into separate falcon tubes for each iCMs, iCFs, and HCAECs.

8. Centrifuge the falcon tube containing HCAECs at 2000 RPM for 4 min and the falcon tubes containing iCMs and iCFs at 300 × *g* for 3 min to form a pellet.
9. Carefully remove supernatant, leaving pellet intact. Resuspend pellets in 1 mL culture medium on ice and perform cell count.

3.2 VCS Formation and Maintenance

The following protocol yields one 96-well plate of CS.

1. Each well of the 96-well Hanging Drop Plate must contain 6000 iCMs, 3000 HCAECs, and 3000 iCFs (i.e., 2:1:1 ratio) in 40 μL of CS medium (*see* **Note 2**). Thus, for a 96-well plate, co-culture 5.76×10^5 iCMs together with 2.88×10^5 HCAECs and iCFs into a falcon tube. Resuspend HCAECs, iCMs, and iCFs in 4 mL VCS medium (*see* **Note 3**), gently mixing by pipetting up and down to achieve a homogenous cell suspension.
2. Pipette 40 μL of cell suspension into each well of the 96-well Hanging Drop Plate. Pipette 1.5 mL sterile 1× PBS in each side of the channel around the Hanging Drop Plate to prevent CS drying out. Incubate Hanging Drop Plates at 37 °C.
3. Check formation of VCS daily and add 5 μL of CS medium every 2 days until cells have aggregated into a 3D CS (*see* **Note 4**) (refer to Fig. 2).

3.3 VCS Isolation from Hanging Drop Culture

1. Taking sterile stainless scissors, a P1000 pipette, and 1000 μL pipette tips, carefully cut ~0.5 cm of the pipette tip off. This will increase the pipette tip diameter and prevent damage to the VCSs as they are taken up by the pipette.
2. Carefully and slowly aspirate the contents of each hanging drop culture well and transfer to a falcon tube on ice by pooling together VCSs (*see* **Note 5**).
3. Centrifuge at 300 × *g* for 2 min to form a pellet. The pellet must be stored on ice until use (*see* **Note 6**).

3.4 Fixation and Immunolabeling for Microscopy of CSs

1. Remove supernatant and fix VCSs in 4% paraformaldehyde for 1 h at room temperature.
2. Remove fixative and wash three times with PBSA for 10 min each wash.
3. Add 500 μL of blocking solution for 30 min at room temperature.
4. Incubate VCS with primary antibodies (10 μg/mL) diluted in 150 μL blocking solution for 1.5–2 h at room temperature or overnight at 4 °C (*see* **Note 7**).
5. Wash three times for 15 min with PBSA at room temperature on a rocking plate (medium speed).

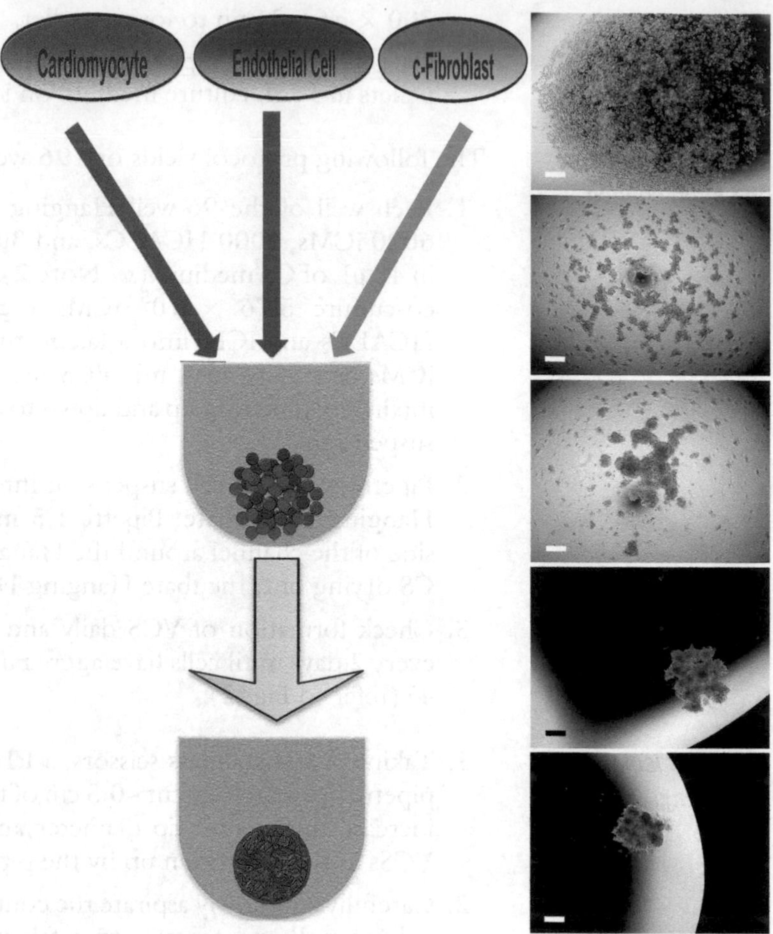

Fig. 2 Generation of VCS in hanging drop culture. Representative bright-field images of human iCM, iCF, and ECs co-cultured in hanging drop after 0, 12, 24, 48, and 96 h. Scale bars: 100 mm [7]

6. Incubate with secondary antibody (10 μg/mL) diluted in 150 μL blocking solution and Hoechst DNA stain for 1.5 h at room temperature or overnight at 4 °C.
7. Wash again three times for 15 min with PBSA at room temperature.
8. Transfer stained VCSs with a previously cut 1000 μL pipette tip to either a glass-bottom plate or Petri dish (*see* **Note 8**).
9. Use plate/Petri dish containing stained VCSs with either a conventional epifluorescent or a confocal microscope for imaging. *Z*-stacks acquired using a confocal microscope can either be collapsed into a single focal plane or used as single images to look through the 3D culture (Fig. 3).

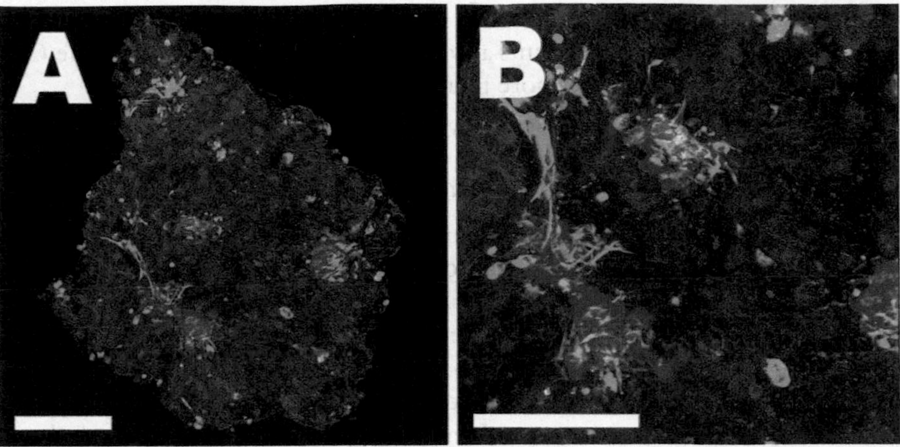

Fig. 3 Sorting of cardiomyocytes, endothelial cells, and fibroblasts within cardiac spheroids. PECAM-positive HCAECs (blue) form a microvascular network, vimentin-positive iCFs (green) sort both on the surface and within the VCS, and Cardiac Troponin T-positive iCMs (red) were surrounded by both HCAECs and iCFs. Scale bar: 200 mm [7]

4 Notes

1. To collect all cellular material during washing, pipette culture medium down sides of cryovial allowing it to rinse residual cells off the walls.
2. To yield more than one 96-well Hanging Drop Plate of VCS, simply multiply this iCM:HCAEC:iCF ratio and volume by the number of Hanging Drop Plates desired.
3. Although 96 wells containing 40 μL each yields an exact total volume of 3.84 mL, it has been better to make a total volume of 4 mL to ensure enough cell suspension for each well.
4. iPSC-derived VCS should form in approximately 3 days (however they may form faster or slower), in which time the non-adherent hanging drop culture causes cell–cell adhesions and aggregations to form. It is not abnormal for multiple VCS to form in each well and for individual VCS to vary slightly in size. Automation using robotic systems, such as Viaflo ASSIST, may be optimal solution for such variation.
5. Since VCS are "sticky," carefully aspirate no more than 3–4 wells of the hanging drop culture plate at any time without releasing contents into falcon tube. Keep VCSs low in the pipette tip and avoid aspirating high into the tip, as they will become "stuck" to the walls of the pipette tip and lost. It is important to carry this process out carefully and slowly to prevent high suction velocities damaging the CS.

6. For optimal cell survival, store the VCS pellet on ice and use in experiments as soon as possible. Avoid leaving the CS pellet on ice for more than ~5 h.
7. VCS stained with antibodies against cardiac troponin T (cTNT), CD31 (PECAM-1), and vimentin, to stain iCMs, HCAECs, and iCFs, respectively.
8. Alternatively, stained VCSs can be transferred on a regular glass slide. To avoid collapsing of VCSs and therefore losing their 3D structure, several glass spacers can be placed on both sides of glass slide, by gently covering VCSs with mounting media and a glass coverslip. Finally, carefully seal the coverslip with nail polish and let it dry for 30 min in the dark prior their use with a microscope. In this way, VCSs can be imaged also after few days, depending on staining used, if kept at 4 °C.

Acknowledgments

This study was supported by a Postdoctoral Marcus Blackmore Fellowship from the Heart Research Institute and a Kick-Start Grant, a Cardiothoracic Surgery Research Grant Scheme and a CDIP Industry & Community Engagement Fund 2017 from the University of Sydney to CG, by an NHMRC Project Grant (APP1129685) to GF and CG, and by a Roche Post-doctoral Fellowship to MC. We would like to thank Dr. John Russell Brereton (Royal North Shore Hospital, Sydney) for his support, Dr. Christine Chuang (University of Copenhagen) for help with the ECM studies, and Dr. Louise Cole (University of Sydney) for their assistance with confocal imaging.

References

1. Gentile C (2016) Filling the gaps between the in vivo and in vitro microenvironment: engineering of spheroids for stem cell technology. Curr Stem Cell Res Ther 11:652–665
2. Fennema E, Rivron N, Rouwkema J et al (2013) Spheroid culture as a tool for creating 3D complex tissues. Trends Biotechnol 31:108–115. https://doi.org/10.1016/j.tibtech.2012.12.003
3. Hakkinen KM, Harunaga JS, Doyle AD, Yamada KM (2011) Direct comparisons of the morphology, migration, cell adhesions, and actin cytoskeleton of fibroblasts in four different three-dimensional extracellular matrices. Tissue Eng A 17:713–724. https://doi.org/10.1089/ten.tea.2010.0273
4. Edmondson R, Broglie JJ, Adcock AF, Yang L (2014) Three-dimensional cell culture systems and their applications in drug discovery and cell-based biosensors. Assay Drug Dev Technol 12:207–218. https://doi.org/10.1089/adt.2014.573
5. Zuppinger C (2016) 3D culture for cardiac cells. Mol Cell Res 1863:1873–1881. https://doi.org/10.1016/j.bbamcr.2015.11.036
6. Baker BM, Chen CS (2012) Deconstructing the third dimension—how 3D culture microenvironments alter cellular cues. J Cell Sci 125:3015–3024
7. Polonchuk L, Chabria M, Badi L et al (2017) Cardiac spheroids as promising in vitro models to study the human heart microenvironment. Sci Rep 7:1–12. https://doi.org/10.1038/s41598-017-06385-8
8. Figtree GA, Bubb KJ, Tang O et al (2017) Vascularized cardiac spheroids as novel 3D

in vitro models to study cardiac fibrosis. Cells Tissues Organs 204:191–198. https://doi.org/10.1159/000477436

9. Benam KH, Dauth S, Hassell B et al (2015) Engineered in vitro disease models. Annu Rev Pathol 10:195–262. https://doi.org/10.1146/annurev-pathol-012414-040418
10. Fitzgerald KA, Malhotra M, Curtin CM et al (2015) Life in 3D is never flat: 3D models to optimise drug delivery. J Control Release 215:39–54. https://doi.org/10.1016/j.jconrel.2015.07.020
11. Visconti RP, Kasyanov V, Gentile C et al (2010) Towards organ printing: engineering an intra-organ branched vascular tree. Expert Opin Biol Ther 10:409–420. https://doi.org/10.1517/14712590903563352

Methods in Molecular Biology (2019) 2002: 61–73
DOI 10.1007/7651_2018_179

Published online: 10 November 2018

Isolation, Propagation, and Clonogenicity of Intestinal Stem Cells

Prashanthi Ramesh, Aleksandar Buryanov Kirov, David Johannes Huels, and Jan Paul Medema

Abstract

Intestinal stem cell research has greatly aided our understanding of the biology of intestinal self-renewal but has also shed light on the role of cancer stem cells (CSCs) in carcinogenesis, cancer growth, and dissemination. With new possibilities for CSC targeting, there is a need to have established techniques for quantifying (cancer) stem cell clonogenicity, particularly in organoid cultures. Here, we describe a detailed methodology for the isolation and expansion of mouse intestinal crypts from three different locations—the colon, proximal, and distal small intestine. In addition, we describe techniques that allow the measurement of stem cell clonogenicity and its manipulation using two approaches—organoid counting and immunohistochemistry.

Keywords Cancer stem cells, Clonogenicity, Colorectal cancer, Immunohistochemistry, Isolation, Organoids

1 Introduction

Multi-region sequencing of tumors reveals the presence of different subpopulations that constitute the heterogeneity observed in some cancers [1]. One of the major contributors to this intra-tumor heterogeneity is the presence of cancer stem cells (CSCs) at the apex of the hierarchical organization within tumors [1]. In colon cancer, CSCs have been identified and extensively studied with the help of specific markers (Lgr5, CD133, CD44, etc.) and are suggested to be responsible for tumor maintenance and propagation upon xenotransplantation [2, 3]. However, the nature, plasticity, and microenvironmental regulation of CSCs still remains a topic of much debate.

Several studies have suggested CSCs to be therapy resistant, thereby contributing to disease relapse and metastasis [1, 4, 5]. This understanding has fueled attempts to identify treatment opportunities to specifically target CSCs. This creates a need for established culture techniques where stem cells and their differentiated progeny can be faithfully represented in vitro. Crypts isolated from mouse and human intestines can be cultured in a 3D Matrigel matrix allowing stem cells to self-organize into mini-gut structures with crypt–villus physiology [6]. These so-called organoids consist of Lgr5+ stem cells and differentiated enterocytes and secretory cells such as Goblet and Paneth cells [6]. Culturing these stem cells in the presence of Wnt amplifier R-spondin, BMP inhibitor Noggin, and EGF allows indefinite maintenance of organoids in vitro and transplantation in vivo [6, 7].

In this chapter, we provide concise methods to study stem cells in crypts derived from genetically modified mouse models. In addition to the extraction of small intestine organoids, we also describe a method for the extraction and culture of mouse colon organoids. 3D culture in Matrigel requires adjustment of usual 2D culture protocols and assays for assessing stem cell sensitivities. In this regard, we outline protocols that allow measurement of treatment-induced changes in clonogenicity and stem cell-specific cell death. Unlike cell lines, seeding organoids is a challenge as the growing structures are often unequally divided between wells, which is further complicated by the fact that only a subset of the cells has the capacity to form new organoids. This experimental variation makes quantification of treatment effects unreliable, which can be overcome by seeding organoids as single cells. However, outgrowth of organoids from single cells is not always successful and requires more time. Here, we describe a technique that allows measurement of organoid clonogenicity without having to grow them out from single cells.

Next to the direct measurement of clonogenicity, immunohistochemical analysis of organoids following treatment can provide insights into population-specific treatment efficacy and changes in cellular states or compositions. Herein, we provide a simple method that allows embedding of whole organoids into paraffin for sectioning and staining. These techniques described can be easily translated to human intestinal and other stem cell-derived organoid cultures.

2 Materials

2.1 General Equipment and Material List

1. Laminar flow cabinet certified for handling biosafety level 2 (ESL-2) specimens
2. Humidified tissue culture incubator capable for maintaining 37 °C and 5% CO_2 atmosphere
3. Low-speed centrifuge (e.g., Hettich Rotanta 460)
4. Inverted and phase contrast light microscopes equipped with 10×, 20×, and 40× magnification
5. 4 °C Cold room
6. Pipet-Aid

7. Tally counter
8. Vortex
9. Sterile disposables plastics: 100 μl filter tips, 1000 μl long-reach pipette tips (VWR), pipettes 1-, 2-, 5-, 10-, and 25-ml volume, 15 and 50 ml polypropylene sterile tubes (Falcon), and 10 cm petri dish
10. Cell culture vessels: 96-,48-, 24-, and 12-well adherent plates (Corning)

2.2 Reagents for Mouse Crypt Isolation, Culture, and Clonogenic Assay

1. Sterile surgical tools for dissection (scissors and tweezers)
2. Fresh intestine derived from mice
3. Sterile disposable plastics: 10 ml syringe, 15 and 50 ml polypropylene sterile tubes (Falcon), 10 cm tissue culture dishes, 96-,48-, 24-, and 12-well adherent plates, 70 μm nylon mesh filter, and glass coverslips
4. Phosphate-buffered saline (PBS) without Ca^{2+} and Mg^{2+}
5. Penicillin–streptomycin (Thermo Fischer Scientific catalog # 15140122)
6. 0.5 M EDTA (Thermo Fischer Scientific catalog# 15575-020)
7. Advanced DMEM/F12 (ADF) medium (Thermo Fischer Scientific catalog #12634-010) (*see* **Note 1**)
8. Growth Factor Reduced Matrigel (Corning) (*see* **Note 2**)
9. *See* Table 1 for crypt culture medium

2.3 Reagents for Paraffin Embedding of Organoids

1. Phosphate-buffered saline (PBS)
2. 4% Paraformaldehyde (PFA)
3. 100 and 70% Ethanol
4. Hematoxylin
5. Xylene
6. Tissue mold (stainless steel)
7. Tissue cassette
8. Paraffin and paraffin dispenser
9. Hot plate (for melting paraffin at 60 °C)
10. Cold plate (for cooling paraffin at 4 °C)
11. Disposable plastics: 12- and 24-well plates, and transfer pipettes

Table 1
Overview of reagents for mouse crypt culture medium

Component	Supplier	Stock solution	Working dilution	Storage
Mouse crypt culture medium				
Advanced DMEM/F12 (ADF medium)	Thermo Fischer Scientific			4 °C
B27 supplement	Thermo Fischer Scientific	50×	1 in 50	−20 °C
N2 supplement	Thermo Fischer Scientific	100×	1 in 100	−20 °C
GlutaMAX-I	Thermo Fischer Scientific	100× (200 mM)	1 in 100	RT[a]
Hepes	Thermo Fischer Scientific	1 M	1 in 100	4 °C
N-acetyl cysteine	Sigma-Aldrich	500 mM	1 in 500	4 °C
Antimycotic/antibiotic	Thermo Fischer Scientific	100×	1 in 100	−20 °C
Mouse EGF	Tebu-BIO	500 μg/ml	1 in 10,000	−20 °C
Noggin conditioned medium[b]			10%	−20 °C
R-spondin conditioned medium[b]			20%	−20 °C
Rock inhibitor[c]	Sigma-Aldrich	10 mM in H_2O	1 in 1000	−20 °C
CHIR-99021[d]	Tebu-BIO	10 mM in DMSO	1 in 3333.3	−20 °C

[a] *RT* room temperature
[b] *See* **Note 3**
[c] *See* **Note 4**
[d] *See* **Note 5**

3 Methods

3.1 Isolation of Crypts from Mouse Proximal and Distal Small Intestine and Colon

Unless otherwise stated, all steps are carried out at room temperature in sterile conditions and the PBS is always supplemented with 50 U/ml penicillin and 50 μg/ml streptomycin.

1. Carefully remove the small intestine from the mouse (*see* **Note 6**) and place it on a clean flat surface (*see* **Note 7**).
2. Using scissors, cut the small intestine into two pieces: small intestinal proximal (SIP) and small intestinal distal (SID) (*see* **Note 8**; *see* Fig. 1). Place the two pieces into a new 10-cm dish.
3. In a separate 10-cm dish, cut and place the colon (*see* **Note 9**).

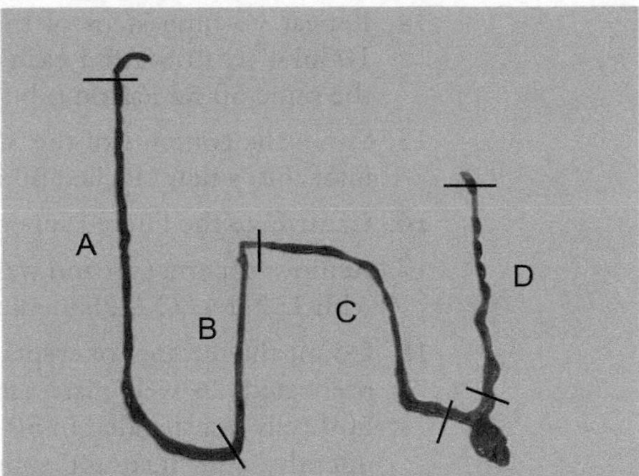

Fig. 1 Dissection of the mouse intestine. Region A from the stomach border to the beginning of the Jejunum is the proximal small intestine (SIP). Region B denotes the medial small intestine and Region C from the ileum to the colon is the distal small intestine (SID). Region D from the caecum to the anus is the colon. The lines indicate the points of division

4. Flush all three pieces with 4 °C PBS to remove any remaining feces (*see* **Note 10**).
5. Carefully cut the intestinal pieces longitudinally open with sterile scissors (*see* **Note 11**) and open up the intestine so that the villi face upwards.
6. Gently scrape the villi off using a glass cover slip (*see* **Note 12**).
7. Cut the proximal and distal intestine into small pieces of approximately 0.5 cm and transfer to two separate 50 ml tubes containing approximately 10 ml 4 °C PBS. From this step onwards, the intestinal pieces should be kept on ice.
8. Cut the colon longitudinally and then into small pieces of approximately 0.5 cm. Then, place in a new 50-ml tube with 10 ml 4 °C PBS.
9. Perform the following steps separately for the SIP, SID, and colon pieces.
10. Wash the intestinal and colon pieces about 5–10 times with 4 °C PBS or until the supernatant is clear (*see* **Note 13**).
11. Resuspend the SIP, SID, and colon in 25 ml of, respectively, 2 mM, 4 mM, and 25 mM EDTA (*see* **Note 14**).
12. Place in cold room (4 °C) on a roller for 30 min.
13. Shake vigorously and collect the supernatant in a new 50-ml tube. Keep the supernatant aside on ice for the remaining steps (*see* **Note 15**).

14. Repeat washing steps of the pieces two to four times using 10 ml 4 °C PBS. After each wash, collect the supernatant into the same 50 ml Falcon tube (*see* **Note 16**).
15. Strain the contents of the 50 ml tube through a 70-μm mesh filter into a new labelled 50 ml tube.
16. Centrifuge the filtered supernatant at 600 × *g* for 5 min.
17. Remove supernatant and wash the cell pellets one to two times with 15 ml 4 °C ADF medium (*see* **Note 17**).
18. Count the number of crypts per isolation and in each well of a preheated 24-well plate (*see* **Note 18**) seed approximately 500 cells resuspended in 50 μl Matrigel (*see* **Note 19**). Subsequently, add medium supplemented with all the growth factors.
19. After 2–3 days, the crypts start to bud (*see* Fig. 2). Refresh the medium every 2–3 days and passage them on a weekly basis (*see* **Note 20**).

3.2 Passaging of Mouse Crypts from Mouse Colon, Proximal and Distal Small Intestine

All steps are performed in sterile conditions.

1. Remove medium from the well.
2. Using a long-reach pipette tip (p1000), pipette 1 ml 4 °C ADF into the well and collect the crypts by disrupting the Matrigel.
3. Pipette the crypts into a labelled 15-ml tube and disrupt vigorously by pipetting up and down (*see* **Note 21**).
4. Add 2 ml 4 °C PBS to the tube and centrifuge at 600 × *g* for 3 min (*see* **Note 22**).
5. Remove as much supernatant as possible and resuspend the cells in Matrigel (*see* **Note 19**). The split ratio for the crypts cells is normally 1:3 (passage one well to three).
6. Allow Matrigel to solidify and then add medium with all the required growth factors (*see* Fig. 3).

3.3 Clonogenic Assay

1. Collect and disrupt organoids in a 15-ml tube as done while passaging.
2. Seed the disrupted organoids into the required number of wells of a 48-well plate (*see* **Note 23**).
3. After giving the organoid structures the time to grow out (usually 3–4 days), count the number of crypts in each well (*see* **Note 24**) and treat them in triplicate.
4. After the desired treatment time, collect each well of organoids into labelled 15 ml tubes and centrifuge at 600 × *g* for 3 min.
5. Remove the supernatant and seed each treatment condition of organoids to a new well of a preheated 24-well plate in 50 μl Matrigel (*see* **Note 25**).

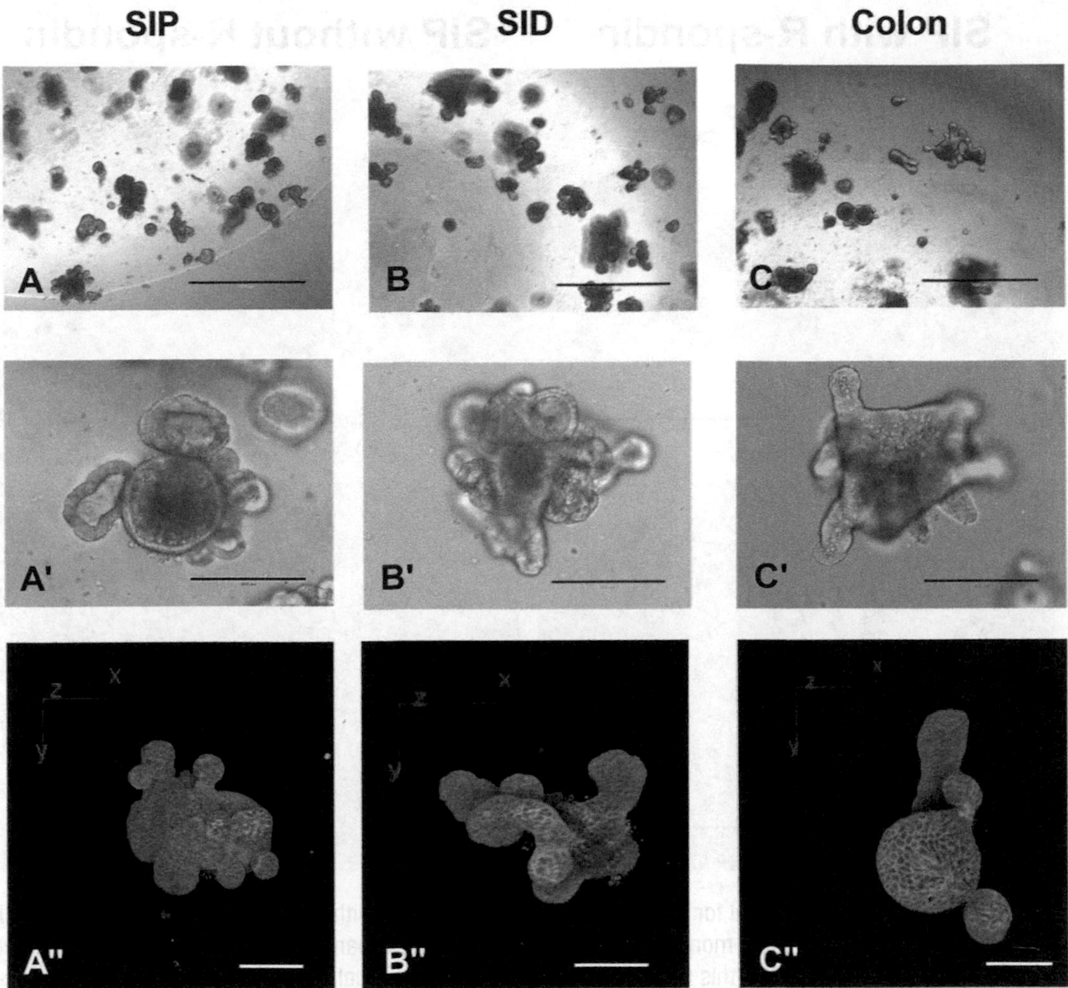

Fig. 2 Phase contrast images of mouse organoids isolated from: (**A** and **A′**) small intestine proximal, (**B** and **B′**) small intestine distal, and (**C** and **C′**) colon. Top row images taken at 4× (scale bars indicate 1000 μm) and middle row at 20× (scale bars indicate 200 μm). 3D images assembled from *z*-stacks taken with confocal microscope of: (**A″**) small intestine proximal, (**B″**) small intestine distal, and (**C″**) colon organoids stained for Actin with Actin-488 (white) and nuclei with Hoechst (blue) (scale bar indicates 100 μm)

6. After 3–4 days, count the number of organoids that grow out for each condition and replicate. By dividing the number of organoids growing out by the number of organoids originally present in the matched 48-well plate, we can estimate the clonogenic capacity of the organoids and also assess if a treatment positively or negatively affects the clonogenic population (*see* Fig. 4).

3.4 Paraffin Embedding of Organoids for IHC

1. Seed organoids in a 12-well plate (*see* **Note 26**).
2. After organoids have grown out, take off the medium and wash the wells twice with PBS without disrupting the Matrigel (*see* **Note 27**).

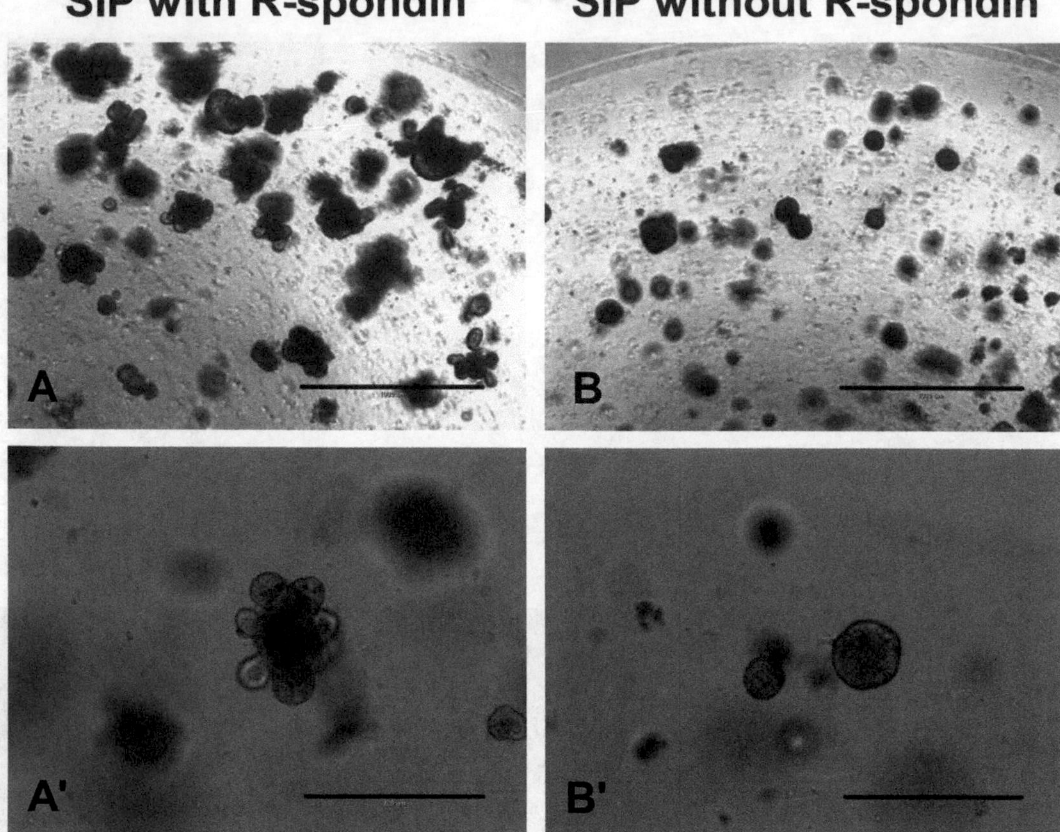

Fig. 3 Growth factors are essential for organoid growth. When cultured with all the required growth factors, SIP organoids maintain their budding morphology (**A** and **A′**) and can be expanded over passages. Cultured in the absence of R-spondin (**B** and **B′**), this morphology is lost and they eventually die. Top row images taken at 4× magnification (scale bars indicate 1000 μm) and bottom row at 10× magnification (scale bars indicate 400 μm)

3. Add 4% paraformaldehyde and fix at 4 °C overnight.
4. Bring the plate to room temperature (RT) for 30 min and remove the paraformaldehyde. Add 70% ethanol and incubate at RT for 30 min (*see* **Note 28**).
5. Remove 70% ethanol and add 100% ethanol containing hematoxylin (1:25 dilution). Incubate at RT for 10–20 min.
6. When the structures start to become colored, take off the hematoxylin–ethanol and gently scrape off the Matrigel and transfer the pieces to a new labelled 24-well plate (*see* **Note 29**).
7. Incubate in hematoxylin–ethanol for another 10 min at RT.
8. Remove the hematoxylin–ethanol, add 500 μl 100% ethanol, and incubate at RT for 20–30 min. Repeat this step and incubate in 100% ethanol for another 20–30 min.

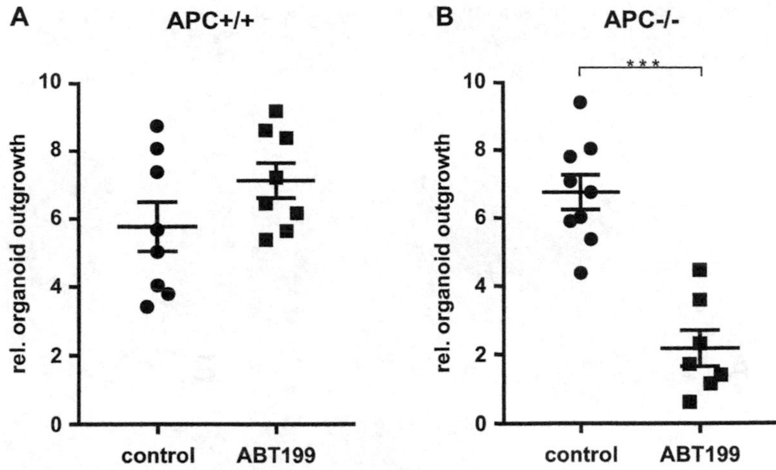

Fig. 4 The clonogenic assay as a reliable readout of stem cell dependencies. SIP organoids obtained from: (**a**) Lgr5.Apc+/+ and (**b**) Lgr5. Apc−/− mice were treated for 72 h with BH3 mimetic ABT-199 (1 μM). Graphs depict the relative outgrowth of organoids counted 3 days after passaging. APC mutant organoids are sensitive to ABT-199 as shown previously [8]

9. Transfer the Matrigel pieces to a metal tissue mold with a labelled tissue cassette.
10. Then, remove the 100% ethanol and add 500 μl xylene in the fume hood, incubate at RT for 20–30 min. Repeat this step and incubate in xylene for another 20–30 min.
11. Remove the xylene, add liquid paraffin to the metal tissue mold, and incubate at 60 °C for 20–30 min.
12. Remove the paraffin using transfer pipettes (*see* **Note 30**) and add fresh paraffin, incubate at 60 °C for another 20–30 min.
13. Remove the paraffin using transfer pipettes and add fresh paraffin. Move the mold to the cold plate (*see* **Note 31**), place the tissue cassette on top, and add some more paraffin to seal the mold. Let cool until the surface has solidified and store the block in −20 °C.
14. Slides can be prepared from this block at 4 μm thickness and stained as required (*see* Fig. 5).

4 Notes

1. For the washing steps in the adenoma isolation procedure, ADF medium is used without any additional factors.
2. To shorten thawing time, we aliquot the Matrigel in 1 ml. Matrigel is thawed on ice at 4 °C overnight or it can be placed on ice approximately 1 h before use.

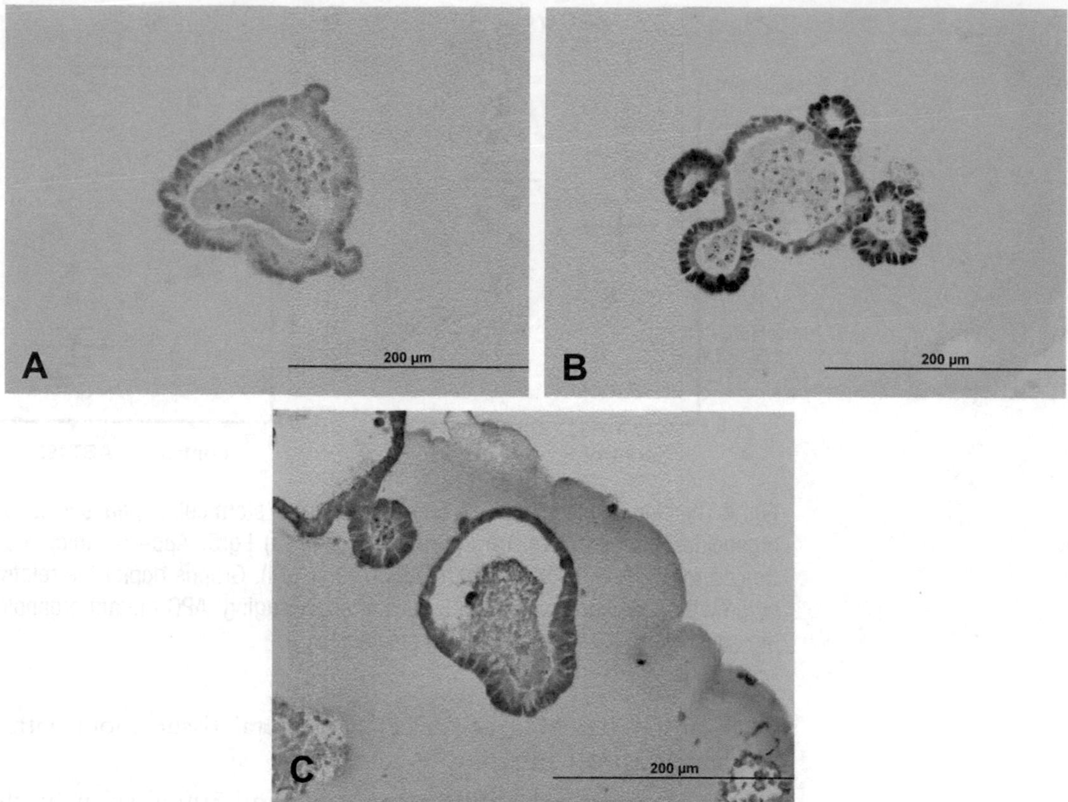

Fig. 5 Immunohistochemical staining of organoids. Paraffin-embedded organoids can be cut and stained for various markers. Depicted here are SIP organoids stained for Alcian blue and Nuclear Red (**a**), Ki67 (**b**) and EPCAM (**c**). Images taken at 20× (scale bars indicate 200 μm)

3. We prepare our own R-spondin and Noggin conditioned medium.
4. Rock inhibitor is used to prevent anoikis and is only used when seeding freshly extracted crypts or when organoids are seeded single cell. After the first passage, there is no need to use this anymore.
5. CHIR-99021 is used only for the culture of colon-derived organoids. Importantly, it is only necessary to use it immediately after passaging of the organoids and should be taken off after 2–3 days in culture when refreshing the medium.
6. The easiest way to remove the intestine is to cut the junction between the stomach and the esophagus. Grab the stomach with forceps, gently lift and pull, as the stomach rises cut away the tissue attached to the intestine, without damaging the intestine. In this manner, both the intestine and colon should untangle, making them easy to remove.

7. We find it easiest to use round 15 cm tissue culture dishes, as they are large enough to easily accommodate the mouse intestine and colon.
8. The proximal small intestine (SIP) refers to the distance from the stomach border with the duodenum to the beginning of the jejunum. The distal small intestine (SID) refers to the distance from the beginning of the ileum to the cecum. Roughly speaking, the SIP is the first 6 cm of intestine and SID is the last 5 cm.
9. The colon refers to the region from the caecum to the anus.
10. For the flushing, the intestine/colon needs to be rinsed inside and out with 4 °C PBS. We use a 0.5 × 16-mm needle attached to a 10-ml syringe to flush the intestine. The stomach and caecum first need to be removed, by cutting at the attachment sites with the duodenum and colon, respectively. The opening of the intestine or colon is then placed over the entry point of the needle and slowly the intestine is filled with PBS. It is important not to fill the intestine too quickly but rather allow the PBS to pass at a passive rate. By lifting the small intestine with the needle, gravity can help the PBS to pass at a passive rate.
11. By segmenting the intestine into three sections, the intestine is easier to cut open longitudinally.
12. The best way to do this is to place the glass cover slip close to where you are holding the beginning of the pieces. Gently press with the glass cover slip and slide towards the opposite end. A white residue should form at the edge of the cover slip, which contains the scraped off villi. This step is not necessary for the colon as it does not contain villi.
13. The washing is performed in order to remove the remaining debris. Antibiotics are included in the PBS in order to avoid bacterial contamination. When washing, shake the 50 ml tube vigorously for 5 s, allow adenoma pieces to settle, and discard supernatant. Repeat this procedure by adding 10 ml 4 °C PBS.
14. The higher concentrations of EDTA improve crypt yields from the SID and colon.
15. You may want to discard this supernatant as it usually does not contain most of the crypts.
16. At this point, you should see the crypts floating around in the PBS after shaking. If you do not see this, incubate in EDTA for a longer time.
17. Pellet volumes generally range from 100 to 500 μl. If the pellet is small (<100 μl), wash once to avoid excessive cell loss. Extractions from SID often result in smaller pellets than SIP and colon.

18. The 24-well plate is preheated for at least 1 h at 37 °C. This helps the solidification of the Matrigel upon plating.
19. For suspension in Matrigel, the cells need to be spun down (3 min, 600 × *g*) and the supernatant removed. Using 50 μl of Matrigel per well, resuspend the cells and drop the Matrigel in the center of the well. The plate is then placed at 37 °C for 10–20 min after which 500 μl of medium per well is added. Avoid making bubbles when resuspending the cells, as this can affect the visualization of structures in the Matrigel.
20. The frequency of passaging is dependent on the rate of crypt growth. Optimally, they should be passaged just before they become overly large, dense, or filled with debris.
21. It is important to mechanically disrupt the structures as much as possible by pipetting up and down at least ten times. This can be done by placing the tip onto the plastic while pipetting in order to increase the pressure on the organoids, thereby dissociating them into smaller fragments. This helps to release them from the Matrigel and make them smaller, allowing further expansion.
22. After centrifugation, the organoid fragments can be seen at the bottom of the tube. A separate transparent layer of Matrigel can be seen above the cells. The Matrigel can easily be aspired and removed.
23. It is best to seed the disrupted organoids at a density that is easy to count. We usually seed around 20–40 organoids per well. This can be achieved by either counting the number of fragments in the tube after isolation, or optimization and standardization of the dilution you need depending on the amount of organoids you start with.
24. It is important to keep track of the number of organoids in each well prior to treatment as this will allow normalization of the number of organoids growing out upon reseeding.
25. The split ratio here is 1:2 to ensure that the organoids have enough space to grow out and that counting does not become too difficult.
26. Organoids are seeded in 100 μl Matrigel into a 12-well plate in order to have bigger pieces with more organoid structures per paraffin block.
27. Washing and fixing can also be done after any desired treatment conditions.
28. At this point, the plate can be stored in 70% ethanol at 4 °C for another week before embedding.
29. Using a p1000 tip, scrape off the Matrigel droplet gently from the sides, trying as much as possible to maintain it in one or

two pieces. Bigger pieces are easier to embed and also avoid loss of organoids during washes.

30. Cut the transfer pipettes at the bottom to make it easier to remove the paraffin.
31. It helps to hold the pieces down with tweezers when placing the mold onto the cold spot as this ensures that the organoids are in approximately the same plane of the block.

Acknowledgments

This work was supported by Oncode Institute, Transcan-2 grant Tactic and Dutch Cancer Society (KWF) Grants UvA2015-7587 and 10150. D.J. Huels was supported by an EMBO long-term fellowship ALTF 1102-2017.

References

1. Nassar D, Blanpain C (2016) Cancer stem cells: basic concepts and therapeutic implications. Annu Rev Pathol 11:47–76
2. Medema JP (2013) Cancer stem cells: the challenges ahead. Nat Cell Biol 15:338–344
3. Vermeulen L, Todaro M, de Sousa MF, Sprick MR, Kemper K et al (2008) Single-cell cloning of colon cancer stem cells reveals a multi-lineage differentiation capacity. Proc Natl Acad Sci U S A 105:13427–13432
4. Oskarsson T, Batlle E, Massagué J (2014) Metastatic stem cells: sources, niches, and vital pathways. Cell Stem Cell 14(3):306–321
5. Colak S, Zimberlin CD, Fessler E, Hogdal L, Prasetyanti PR et al (2014) Decreased mitochondrial priming determines chemoresistance of colon cancer stem cells. Cell Death Differ 21 (7):1170–1177
6. Sato T, Vries RG, Snippert HJ, van de Wetering M, Barker N et al (2009) Single Lgr5 stem cells build crypt-villus structures in vitro without a mesenchymal niche. Nature 459:262–265
7. Yui S, Nakamura T, Sato T, Nemoto Y, Mizutani T et al (2012) Functional engraftment of colon epithelium expanded in vitro from a single adult Lgr5+ stem cell. Nat Med 18:618–623
8. Van Der Heijden M, Zimberlin CD, Nicholson AM, Colak S, Kemp R et al (2016) Bcl-2 is a critical mediator of intestinal transformation. Nat Commun 7:10916

Methods in Molecular Biology (2019) 2002: 75–85
DOI 10.1007/7651_2018_183

Published online: 23 September 2018

Isolation of Extracellular Vesicles from Subventricular Zone Neural Stem Cells

Mary C. Morton, Victoria N. Neckles, and David M. Feliciano

Abstract

The neonatal subventricular zone (SVZ) is a neurogenic niche that contains neural stem cells (NSCs). NSCs release particles called extracellular vesicles (EVs) that contain biological material. EVs are transferred to cells, including immune cells in the brain called microglia. A standard approach to identify EV functions is to isolate and transplant EVs. Here, a detailed protocol is provided that will allow one to culture neonatal SVZ NSCs and to isolate, label, and transplant EVs. The protocol will permit careful and thorough examination of EVs in a wide range of physiological and pathophysiological conditions.

Keywords Exosome, Extracellular vesicles, Microglia, Microvesicle, Neural stem cell, Subventricular zone

1 Introduction

Extracellular vesicles (EVs) are particles that range in size from 50 to 350 nm and are categorized by their mechanism of biogenesis [1]. Exosomes are EVs that are generated upon fusion of the cell membrane with multivesicular bodies that causes the release of intraluminal vesicles into the extracellular space [2]. Microvesicles are generated by scission of outward budding protuberances of the cell membrane [2]. These two categories of EVs are membrane encapsulated. Non-membranous EVs called exomeres are proteinaceous particles comprised of metabolic, translational, and coagulation regulating proteins [3]. EVs can be transferred from donor to recipient cells thereby serving as a novel mode of intercellular communication [4, 5]. One mechanism by which this occurs is the transfer of nucleic acids which regulate viral nucleic acid sensing pathways [6]. In other cases, metabolic enzymes may be transferred to cells in a paracrine-like manner [7].

Various methods are used for isolating EVs. Polymer-based isolation utilizes polyethylene glycol to isolate a high yield of membraneous EVs but contain lipoprotein components [8–10]. Ultracentrifugation can be used to isolate a mixture of EVs but often lack homogeneity in EV types purified which can lead to inconsistencies that hamper reproducibility [11, 12]. Therefore, cleanup steps are recommended to further purify EV types following ultracentrifugation. EV pellets are resuspended in high molarity (M) sucrose and centrifuged on a sucrose cushion. EVs are then placed onto a sucrose gradient. It is recommended that the finished gradient containing EVs is centrifuged at a very high speed for at least 16 h to ensure proper EV sedimentation [13]. Sucrose density gradient fractionation relies on inherent differences of buoyancy of EVs for isolation of enriched fractions. The buoyancy may however rely on the source, content, and physiological state for which the EVs are isolated. Microvesicles and exosomes are then collected as fractions at a sucrose concentration of 1.1–1.9 g/mL, but EV types overlap and are not separated completely [13, 14]. Microvesicles often elute in fractions 2–4 and exosomes from fractions 5–9. Sucrose density gradients have their own limitations including structural compromise, EV fusion, and vesicle rupture [15]. Currently, there is no universal standard for EV isolation.

The subventricular zone (SVZ) is one of the two neurogenic niches in the perinatal brain [16]. Located contiguous with the lateral ventricles, one population of SVZ neural stem cells (NSCs) symmetrically divide as a mode of self-renewal [17]. Whereas a separate, larger population of NSCs give rise to Type C cells that divide three or four times to generate neuroblasts which migrate along the rostral migratory stream to the olfactory bulb where they differentiate into periglomerular or granule interneurons [18–22]. Reduction in the number of SVZ NSCs occurs in correlation in both aging and in continued NSC proliferation [23, 24]. Additionally, the number of SVZ NSCs decreases by ~60% between early postnatal life (0–7 days) and adults at 26 months [25]. The neurogenic zone in the SVZ decrease significantly between 0 and 15 days postnatally (P0–P15), and at P15, the SVZ begins to resemble that of an adult SVZ [26]. SVZ NSCs produce astrocytes and oligodendrocytes, the support cells of the nervous system, and ependymal cells that act as a barrier between the SVZ and the fluid-filled lateral ventricles [18, 27]. The early perinatal SVZ (P0-P1) is comprised mainly of NSCs. Resident central nervous system (CNS) immune cells infiltrate the developing nervous system from the yolk sac at embryonic day 8.5–9.5 [28]. These cells, called microglia, congregate in proliferative zones in the perinatal brain that later in postnatal development migrate out to and populate the cortex [29, 30]. Microglia release factors that regulate NSC proliferation during embryonic development and into perinatal neurogenesis [31–34]. Recent studies have identified SVZ-NSCs as sources of EVs that regulate neighboring cells, including microglia [35–37].

When culturing primary SVZ NSCs, various methods have been described in which their utility is dependent on the goal of the experiment. Primary SVZ NSCs have been cultured through neurosphere assays, adherent monolayer systems, and matrigels [38–41]. Neurosphere assays allow for the examination of differences in cell proliferation and cell potential [42]. Neurospheres themselves are heterogenous clusters of NSCs, progenitor cells, and differentiated cells that, together, generate a stem cell niche with more differentiated cells residing in the center [43–46]. Unlike neurosphere cultures, adherent monolayer systems more closely recapitulate in vivo proliferation and have a more homogenous population of NSCs [42, 47]. The homogeneity of adherent monolayer systems allows for a better interpretation of NSCs rather than NSC niches of neurospheres. Finally, matrigels are used to construct a 3D structure that mimics the extracellular matrix in which NSCs reside in vivo [48]. Different from nonadherent cell culture systems, such as neurospheres, cells cultured in matrigels have a distinct advantage in that they are cultured in a degradable biomaterial which provides structural integrity [49]. Matrigel culturing systems allow for examination of neuro-regeneration and potency [49, 50].

Here, we demonstrate in detail the culturing of the neonatal SVZ, which is enriched in NSCs, the isolation of EVs with the production of a "dirty" fraction, and a subsequent cleanup step, followed by the labeling and transplantation of EVs into the developing perinatal lateral ventricles [42].

2 Materials

2.1 Primary Cell Culture and Microdissection

(a) Mouse laminin

(b) Neurobasal A Complete Culture Media: Neurobasal A media, 2% B27, 1× Glutamax, 50 units/mL penicillin/streptomycin, 20 ng/mL purified mouse receptor-grade epidermal growth factor (EGF), and 20 ng/mL recombinant bovine fibroblast growth factor (FGF-2)

(c) Dissociation buffer: 0.05% trypsin–EDTA, Neurobasal A media

(d) Cascade Biologics trypsin inhibitor

(e) Sterile microdissection kit

(f) Scalpel

2.2 Sucrose Density Gradient Exosome Isolations

(a) Thick wall polycarbonate tube or ultracentrifuge tubes

(b) Sucrose for density gradient: 2.5 M sucrose, 1× d-phosphate-buffered saline solution (dPBS)

(c) Beckman Coulter Optima MAX-XP centrifuge with TLA 100.3 rotor

2.3 Exosome Labeling and Transplantation

(a) Vybrant DiI cell-labeling solution

(b) 10 cm Fire polished borosilicate glass capillary tubes or Hamilton Neuros Syringe

(c) Tabletop centrifuge

3 Methods

3.1 Preparation of Culture Media

1. 24 h prior to culturing cells, prepare laminin-coated wells for adherent monolayer cultures. To prepare culture wells, add 8.5 μg/mL in DiH_2O per well and incubate at 37 °C overnight. Wash wells three times using DiH_2O and use wells immediately.
2. Prepare fresh culture media on the day of dissections by mixing Neurobasal A media with 2% B27, 1× Glutamax, 50 units/mL penicillin/streptomycin, 20 ng/mL purified mouse receptor-grade epidermal growth factor (EGF), and 20 ng/mL recombinant bovine fibroblast growth factor (FGF-2). Warm culture media to 37 °C (*see* **Note 1**).
3. Prepare glass Pasteur pipettes with varying bore sizes by rotating tip of glass pipette over an open flame until desired bore size is obtained. Ideally, "small," "medium," and "large" bore are best suited for culturing (*see* **Note 2**).
4. Prepare 0.05% trypsin–EDTA in Neurobasal A for the dissociation buffer, and 1× trypsin inhibitor (Cascade Biologics). Pre-warm to 37 °C.
5. Sterilize dissection tools (forceps 2×, microdissection scissors, dissection scissors, dissecting microscope, and scalpel) (*see* **Note 3**).

3.2 Harvesting Perinatal Brains and SVZ Microdissections

1. Anesthetize P0–P1 pup by placing it on ice for 5 min or by following the facility's proper guidelines.
2. Decapitate pup using scissors and carefully remove the skull by cutting between the two hemispheres down the midline along the sagittal suture and using forceps to peel back the skin and skull thus exposing the brain (*see* **Note 4**).
3. Carefully remove the brain by using the curved or soft edges of the forceps and place the brain into ice-cold 1× phosphate-buffered saline (PBS) solution under the dissecting scope with the ventral side up.
4. Using the scalpel, remove olfactory bulbs. Cut a coronal section from the brain 1/3 of the distance from the most rostral portion of the cortex, severing the cortex into two pieces. Discard the most caudal piece.

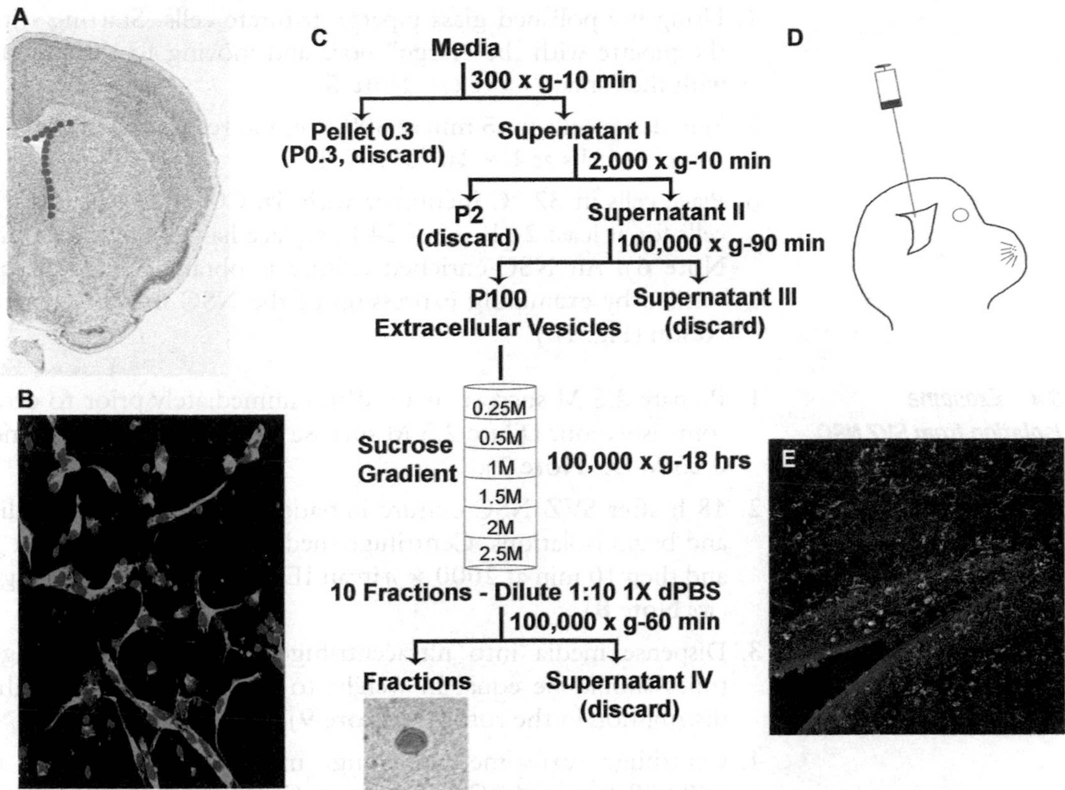

Fig. 1 *Isolated EVs are targeted to microglia.* (**a**) Coronal section of a mouse brain with the SVZ highlighted (purple) dissected region. Image credit: Allen Institute. (**b**) An example of an SVZ culture stained for Nestin (green) and DNA (TO-PRO-3, blue). (**c**) Conditioned media is subjected to a serial centrifugation protocol that produces a "dirty fraction." EVs are further purified on by sucrose gradient fractionation. An example electron micrograph of an NSC exosome. (**d**) Schematic of EV injection into the lateral ventricle of a P0 pup. (**e**) Coronal section showing SVZ after DiI-labeled EVs (red) and stained for the microglia markers Iba1 (blue) and CD11b (green)

5. To microdissect the SVZ, place the rostral portion of the cortex with the cut side facing up. Carefully remove the SVZ contiguous with the dorsal and lateral wall of the lateral ventricles and place dissected SVZ into Neurobasal A incubated on ice (Fig. 1a).

3.3 SVZ Tissue Dissociation

1. Place dissected SVZ into 750 μL dissociation buffer (0.05% trypsin–EDTA in Neurobasal A), gently invert the sample two or three times, and incubate at 37 °C for 7 min.
2. Add 1:1 ratio of 1× Cascade Biologics trypsin inhibitor to dissociation buffer and centrifuge for 5 min at 300 × *g* in Eppendorf centrifuge 5415 D.
3. Aspirate supernatant and resuspend cell pellet in 1000 μL pre-warmed Neurobasal A complete media.

4. Using fire polished glass pipette, triturate cells. Starting with the pipette with the "large" bore and moving to the pipette with the "small" bore (*see* **Note 5**).
5. Spin down cells for 5 min at 300 × *g*, and resuspend in 200 μL and plate cells at 1 × 10^6 cells/mL.
6. Place cells in 37 °C incubator with 5% CO_2. Do not disturb cells for at least 24 h. After 24 h, replace half of the media (*see* **Note 6**). An NSC-enriched culture is obtained and can be verified by examining expression of the NSC marker protein Nestin (Fig. 1b).

3.4 Exosome Isolation from SVZ NSC Culture Media Using Sucrose Density Gradients

1. Prepare 2.5 M sucrose in 1× dPBS immediately prior to exosome isolation. Dilute 2.5 M sucrose to 2.0, 1.5, 1.0, 0.5, and 0.25 M (*see* **Note 7**).
2. 48 h after SVZ NSC culture initiation, collect culture media and begin isolations. Centrifuge media for 10 min at 300 × *g* and then 10 min at 2000 × *g* in an IEC Centra GP8 centrifuge (*see* **Note 8**).
3. Dispense media into ultracentrifuge tubes. Ultracentrifuge tubes should be equal in weight to prevent unequal weight distribution in the rotor (*see* **Note 9**).
4. Centrifuge exosome-containing media for 90 min at 100,000 × *g* at 4 °C in Beckman Coulter Optima MAX-XP centrifuge with TLA 100.3 rotor. This will result in a pellet hereafter referred to as the P100 pellet or dirty fraction (*see* **Note 10**).
5. Resuspend the P100 fraction in 2.5 M sucrose and layer the other sucrose solution in descending order (2.0, 1.5, and 1.0 M, etc.). Take extra caution not to disturb any layer, otherwise the gradient will not form properly. Weigh the completed sucrose gradients and ensure equal weight before moving on to the next centrifuge step (*see* **Note 11**).
6. Place gradients in ultracentrifuge rotor and place the rotor in the ultracentrifuge. Spin gradient for 18 h at 4 °C at 100,000 × *g*.
7. Discard top layer and collect ten (1–10), equal fractions and place each fraction in a separate ultracentrifuge tube. Dilute 1:10 in 1× dPBS and spin each fraction for 1 h at 100,000 × *g* at 4 °C (*see* **Note 12**).
8. Discard supernatant and resuspend each fraction in 30 μL 1× dPBS. Collect exosome fractions 5–8. Exosomes can be stored at −20 °C or immediately labeled and used for transplantation (*see* **Note 13**). A schematic of the EV isolation protocol is included (Fig. 1c).

3.5 DiI Labeling of Exosomes and Transplantation

1. Centrifuge exosomes at 14,000 × *g* for 30 min.
2. From a 1-mM stock solution of DiI-labeling solution, resuspend pellets in 1 μM DiI-labeling solution in 1× dPBS and incubate fractions for 10 min at room temperature and vortex periodically during incubation.
3. Centrifuge fractions at 14,000 (or 100,000) × *g* for 30 min at room temperature.
4. Resuspend fractions in 1× PBS and repeat steps 3.5.2 and 3.5.3 3× times (*see* **Note 14**).
5. Resuspend final pellet of DiI-labeled exosomes in 50 μL 1× dPBS and either store exosomes at −20 °C or move immediately to transplantation.

3.6 Transplantation of Exosomes into Lateral Ventricle of P0 Pups

1. Prior to transplantations, prepare pulled glass pipettes by placing 10 cm fire polished borosilicate glass capillary tubes (O.D.: 1.5 mm, I.D.: 1.1 mm) into a Sutter Instrument Company Model P-97 pipette puller. Place pipettes in a sterile container for transplantations. Alternatively, a Hamilton Neuros Syringe can also be used for injections.
2. Load 1–2 μL of DiI-labeled exosomes into Hamilton Neuros Syringe constructed with a Neuros Adapter.
3. Anesthetize P0 pup on ice for 5 min or until pup is no longer ambulatory. Hold pup in between thumb, index and middle finger (nose should be anchored using thumb). Make sure skin is taut to identify injection site.
4. Identify sagittal suture of the developing skull along the midline of the brain. Place capillary needle or Hamilton Neuros Syringe needle near the rostral portion of the brain at the midline. Move the needle approximately 1 mm laterally in either direction and 0.5 mm caudally and insert needle into the lateral ventricle approximately 2 mm deep. Inject 1–2 μL DiI-labeled exosomes into the lateral ventricles.
5. Place pups on heating pad for 5 min. Once warmed, pup is placed with mother. Sacrifice pups after transplantation, collect, and fix brain. EVs colocalize with microglia (Fig. 1e).

4 Notes

1. Neurobasal A complete media can be stored at 4 °C for 2 weeks. It is recommended that fresh complete media be made for this protocol.
2. It is recommended that three glass Pasteur pipettes with varying bore sizes should be prepared each with decreasing bore size. Before using on SVZ tissue, be sure that liquid can pass through the newly sized bore.

3. All tools should be sterilized prior to tissue dissection. Primary cultured cells are extremely susceptible to contamination. Any steps involving the dissected tissue should be completed under a cell culture hood using 70% ethanol to sterilize when appropriate.
4. When removing the brain, be sure to not cut too deep between the hemispheres or use the sharp edges of the dissection forceps to remove the skull. Either could result in puncturing the brain and damaging the tissue.
5. Media should have a cloudy appearance indicating proper tissue dissociation. Do not triturate cells excessively, 7–10 times total is sufficient.
6. Save media. Do not discard.
7. For each EV isolation procedure, be sure to prepare fresh 2.5 M sucrose solution in 1× dPBS. Addition of antibiotics and antifungals may aid in preventing contamination for storage.
8. Transfer media into new tubes between centrifugation steps. Be sure to not disturb the pellet.
9. When weighing the ultracentrifuge tubes, the tubes should be equal in weight to the 100th decimal place to prevent systemic mechanical breakdown.
10. Prior to ultracentrifugation, ensure that tubes with the sample are equal in weight to the 100th decimal place. Additionally, P100 fractions can be used for postnatal transplantations. P100 fraction contains both exosomes and microvesicles.
11. In some cases, a syringe is most useful to construct the sucrose gradient layers. When constructing the sucrose density gradient, do not add the next layer of sucrose directly onto the previous layer in a destructive manner. Instead, pipette each layer of sucrose down the side of the tube, gently, to ensure that the prior layer is not disrupted. Again, weigh each tube containing the sucrose gradients and equilibrate using 1× dPBS.
12. When collecting fractions, collect from the top down. Be careful not to insert the pipette tip too far into the gradient, thus disrupting lower layers.
13. Exosome pellets can appear as a gray or translucent pellet. Take note of tube orientation in the rotor before resuspending pellet.
14. To ensure no DiI-labeling particles are left in the exosome pellet, repeat the washes in 1× dPBS as many times as needed.

Acknowledgments

David M. Feliciano is supported by grants from the Whitehall Foundation and National Institutes of Health 1R15NS096562.

References

1. Cocucci E, Meldolesi J (2015) Ectosomes and exosomes: shedding the confusion between extracellular vesicles. Trends Cell Biol 25:364–372
2. Théry C, Zitvogel L, Amigorena S (2002) Exosomes: composition, biogenesis and function. Nat Rev Immunol 2:569–579
3. Zhang H, Freitas D, Kim HS, Fabijanic K, Li Z, Chen H, Mark MT, Molina H, Martin AB, Bojmar L, Fang J, Rampersaud S, Hoshino A, Matei I, Kenific CM, Nakajima M, Mutvei AP, Sansone P, Buehring W, Wang H, Jimenez JP, Cohen-Gould L, Paknejad N, Brendel M, Manova-Todorova K, Magalhães A, Ferreira JA, Osório H, Silva AM, Massey A, Cubillos-Ruiz JR, Galletti G, Giannakakou P, Cuervo AM, Blenis J, Schwartz R, Brady MS, Peinado H, Bromberg J, Matsui H, Reis CA, Lyden D (2018) Identification of distinct nanoparticles and subsets of extracellular vesicles by asymmetric flow field-flow fractionation. Nat Cell Biol 20:332–343
4. Ramachandran S, Palanisamy V (2012) Horizontal transfer of RNAs: exosomes as mediators of intercellular communication. Wiley Interdiscip Rev RNA 3:286–293
5. Valadi H, Ekström K, Bossios A, Sjöstrand M, Lee JJ, Lötvall JO (2007) Exosome-mediated transfer of mRNAs and microRNAs is a novel mechanism of genetic exchange between cells. Nat Cell Biol 9:654–659
6. Eckard SC, Rice GI, Fabre A, Badens C, Gray EE, Hartley JL, Crow YJ, Stetson DB (2014) The SKIV2L RNA exosome limits activation of the RIG-I-like receptors. Nat Immunol 15:839–845
7. Iraci N, Gaude E, Leonardi T, Costa ASH, Cossetti C, Peruzzotti-Jametti L, Bernstock JD, Saini HK, Gelati M, Vescovi AL, Bastos C, Faria N, Occhipinti LG, Enright AJ, Frezza C, Pluchino S (2017) Extracellular vesicles are independent metabolic units with asparaginase activity. Nat Chem Biol 13:951–955
8. Vickers KC, Palmisano BT, Shoucri BM, Shamburek RD, Remaley AT (2011) MicroRNAs are transported in plasma and delivered to recipient cells by high-density lipoproteins. Nat Cell Biol 13:423–433
9. Lobb RJ, Becker M, Wen SW, Wong CSF, Wiegmans AP, Leimgruber A, Möller A (2015) Optimized exosome isolation protocol for cell culture supernatant and human plasma. J Extracell vesicles 4:27031
10. Momen-Heravi F, Saha B, Kodys K, Catalano D, Satishchandran A, Szabo G (2015) Increased number of circulating exosomes and their microRNA cargos are potential novel biomarkers in alcoholic hepatitis. J Transl Med 13:261
11. Livshits MA, Khomyakova E, Evtushenko EG, Lazarev VN, Kulemin NA, Semina SE, Generozov EV, Govorun VM, Govorun VM (2015) Isolation of exosomes by differential centrifugation: Theoretical analysis of a commonly used protocol. Sci Rep 5:17319
12. Bobrie A, Colombo M, Krumeich S, Raposo G, Théry C (2012) Diverse subpopulations of vesicles secreted by different intracellular mechanisms are present in exosome preparations obtained by differential ultracentrifugation. J Extracell Vesicles 1:18397
13. Taylor DD, Shah S (2015) Methods of isolating extracellular vesicles impact down-stream analyses of their cargoes. Methods 87:3–10
14. Théry C, Amigorena S, Raposo G, Clayton A (2006) Isolation and characterization of exosomes from cell culture supernatants and biological fluids. Curr Protoc Cell Biol 30:3.22.1–3.22.29
15. Linares R, Tan S, Gounou C, Arraud N, Brisson AR (2015) High-speed centrifugation induces aggregation of extracellular vesicles. J. Extracell. Vesicles. 4:29509
16. Lim DA, Alvarez-Buylla A (2016) The adult ventricular-subventricular zone (V-SVZ) and olfactory bulb (OB) neurogenesis. Cold Spring Harb Perspect Biol 8:a018820
17. Obernier K, Cebrian-Silla A, Thomson M, Parraguez JI, Anderson R, Guinto C, Rodas Rodriguez J, Garcia-Verdugo J-M, Alvarez-Buylla A (2018) Adult neurogenesis is sustained by symmetric self-renewal and differentiation. Cell Stem Cell 22:221–234.e8

18. Bjornsson CS, Apostolopoulou M, Tian Y, Temple S (2015) It takes a village: constructing the neurogenic niche. Dev Cell 32:435–446
19. Luskin MB (1993) Restricted proliferation and migration of postnatally generated neurons derived from the forebrain subventricular zone. Neuron 11:173–189
20. Lois C, Garcia-Verdugo J-M, Alvarez-Buylla A (1996) Chain migration of neuronal precursors. Science 271:978–981
21. Petreanu L, Alvarez-Buylla A (2002) Maturation and death of adult-born olfactory bulb granule neurons: role of olfaction. J Neurosci 22:6106–6113
22. Imayoshi I, Sakamoto M, Ohtsuka T, Takao K, Miyakawa T, Yamaguchi M, Mori K, Ikeda T, Itohara S, Kageyama R (2008) Roles of continuous neurogenesis in the structural and functional integrity of the adult forebrain. Nat Neurosci 11:1153–1161
23. Bouab M, Paliouras GN, Aumont A, Forest-Bérard K, Fernandes KJL (2011) Aging of the subventricular zone neural stem cell niche: evidence for quiescence-associated changes between early and mid-adulthood. Neuroscience 173:135–149
24. Daynac M, Morizur L, Chicheportiche A, Mouthon M-A, Boussin FD (2016) Age-related neurogenesis decline in the subventricular zone is associated with specific cell cycle regulation changes in activated neural stem cells. Sci Rep 6:21505
25. Maslov AY, Barone TA, Plunkett RJ, Pruitt SC (2004) Neural stem cell detection, characterization, and age-related changes in the subventricular zone of mice. J Neurosci 24:1726–1733
26. Tramontin AD, García-Verdugo JM, Lim DA, Alvarez-Buylla A (2003) Postnatal development of radial glia and the ventricular zone (VZ): a continuum of the neural stem cell compartment. Cereb Cortex 13:580–587
27. Mirzadeh Z, Merkle FT, Soriano-Navarro M, Garcia-Verdugo JM, Alvarez-Buylla A (2008) Neural stem cells confer unique pinwheel architecture to the ventricular surface in neurogenic regions of the adult brain. Cell Stem Cell 3:265–278
28. Ginhoux F, Greter M, Leboeuf M, Nandi S, See P, Gokhan S, Mehler MF, Conway SJ, Ng LG, Stanley ER, Samokhvalov IM, Merad M (2010) Fate mapping analysis reveals that adult microglia derive from primitive macrophages. Science 330:841–845
29. Cunningham CL, Martínez-Cerdeño V, Noctor SC (2013) Microglia regulate the number of neural precursor cells in the developing cerebral cortex. J Neurosci 33:4216–4233
30. Swinnen N, Smolders S, Avila A, Notelaers K, Paesen R, Ameloot M, Brône B, Legendre P, Rigo J-M (2013) Complex invasion pattern of the cerebral cortex bymicroglial cells during development of the mouse embryo. Glia 61:150–163
31. Battista D, Ferrari CC, Gage FH, Pitossi FJ (2006) Neurogenic niche modulation by activated microglia: transforming growth factor β increases neurogenesis in the adult dentate gyrus. Eur J Neurosci 23:83–93
32. Zhu P, Hata R, Cao F, Gu F, Hanakawa Y, Hashimoto K, Sakanaka M (2008) Ramified microglial cells promote astrogliogenesis and maintenance of neural stem cells through activation of Stat3 function. FASEB J 22:3866–3877
33. Antony JM, Paquin A, Nutt SL, Kaplan DR, Miller FD (2011) Endogenous microglia regulate development of embryonic cortical precursor cells. J Neurosci Res 89:286–298
34. Snyder EY, Yoon C, Flax JD, Macklis JD (1997) Multipotent neural precursors can differentiate toward replacement of neurons undergoing targeted apoptotic degeneration in adult mouse neocortex. Proc Natl Acad Sci U S A 94:11663–11668
35. Morton MC, Neckles VN, Seluzicki CM, Holmberg JC, Feliciano DM (2018) Neonatal subventricular zone neural stem cells release extracellular vesicles that act as a microglial morphogen. Cell Rep 23:78–89
36. Cossetti C, Iraci N, Mercer TR, Leonardi T, Alpi E, Drago D, Alfaro-Cervello C, Saini HK, Davis MP, Schaeffer J, Vega B, Stefanini M, Zhao C, Muller W, Garcia-Verdugo JM, Mathivanan S, Bachi A, Enright AJ, Mattick JS, Pluchino S (2014) Extracellular vesicles from neural stem cells transfer IFN-γ via Ifngr1 to activate Stat1 signaling in target cells. Mol Cell 56:193–204
37. Asai H, Ikezu S, Tsunoda S, Medalla M, Luebke J, Haydar T, Wolozin B, Butovsky O, Kügler S, Ikezu T (2015) Depletion of microglia and inhibition of exosome synthesis halt tau propagation. Nat Neurosci 18:1584–1593
38. Ray J, Raymon HK, Gage FH (1995) Generation and culturing of precursor cells and neuroblasts from embryonic and adult central nervous system. Methods Enzymol 254:20–37
39. Palmer TD, Ray J, Gage FH (1995) FGF-2-responsive neuronal progenitors reside in proliferative and quiescent regions of the adult rodent brain. Mol Cell Neurosci 6:474–486

40. Reynolds BA, Weiss S (1992) Generation of neurons and astrocytes from isolated cells of the adult mammalian central nervous system. Science 255:1707–1710
41. Reynolds BA, Weiss S (1996) Clonal and population analyses demonstrate that an EGF-responsive mammalian embryonic CNS precursor is a stem cell. Dev Biol 175:1–13
42. Walker TL, Kempermann G (2014) One mouse, two cultures: isolation and culture of adult neural stem cells from the two neurogenic zones of individual mice. J Vis Exp. https://doi.org/10.3791/51225
43. Azari H, Rahman M, Sharififar S, Reynolds BA (2010) Isolation and expansion of the adult mouse neural stem cells using the neurosphere assay. J Vis Exp. https://doi.org/10.3791/2393
44. Bez A, Corsini E, Curti D, Biggiogera M, Colombo A, Nicosia RF, Pagano SF, Parati EA (2003) Neurosphere and neurosphere-forming cells: morphological and ultrastructural characterization. Brain Res 993:18–29
45. Suslov ON, Kukekov VG, Ignatova TN, Steindler DA (2002) Neural stem cell heterogeneity demonstrated by molecular phenotyping of clonal neurospheres. Proc Natl Acad Sci U S A 99:14506–14511
46. Parmar M, Sjöberg A, Björklund A, Kokaia Z (2003) Phenotypic and molecular identity of cells in the adult subventricular zone. in vivo and after expansion in vitro. Mol Cell Neurosci 24:741–752
47. Babu H, Claasen J-H, Kannan S, Rünker AE, Palmer T, Kempermann G (2011) A protocol for isolation and enriched monolayer cultivation of neural precursor cells from mouse dentate gyrus. Front Neurosci 5:89
48. Aligholi H, Rezayat SM, Azari H, Ejtemaei Mehr S, Akbari M, Modarres Mousavi SM, Attari F, Alipour F, Hassanzadeh G, Gorji A (2016) Preparing neural stem/progenitor cells in PuraMatrix hydrogel for transplantation after brain injury in rats: a comparative methodological study. Brain Res 1642:197–208
49. Thonhoff JR, Lou DI, Jordan PM, Zhao X, Wu P (2008) Compatibility of human fetal neural stem cells with hydrogel biomaterials in vitro. Brain Res 1187:42–51
50. Moradi F, Bahktiari M, Joghataei MT, Nobakht M, Soleimani M, Hasanzadeh G, Fallah A, Zarbakhsh S, Hejazian LB, Shirmohammadi M, Maleki F (2012) BD PuraMatrix peptide hydrogel as a culture system for human fetal Schwann cells in spinal cord regeneration. J Neurosci Res 90:2335–2348

Methods in Molecular Biology (2019) 2002: 87–99
DOI 10.1007/7651_2018_186

Published online: 06 September 2018

Reconstruction of Regenerative Stem Cell Niche by Cell Aggregate Engineering

Bing-Dong Sui, Bin Zhu, Cheng-Hu Hu, Pan Zhao, and Yan Jin

Abstract

The niche plays critical roles in regulating functionality and determining regenerative outcomes of stem cells, for which establishment of favorable microenvironments is in demand in translational medicine. In recent years, the cell aggregate technology has shown immense potential to reconstruct a beneficial topical niche for stem cell-mediated regeneration, which has been recognized as a promising concept for high-density stem cell delivery with preservation of the self-produced, tissue-specific extracellular matrix microenvironments. Here, we describe the basic methodology of stem cell aggregate-based niche engineering and quality check indexes prior to application.

Keywords Cell aggregate, Extracellular matrix, Microenvironment, Niche engineering, Stem cells, Tissue regeneration

1 Introduction

As stem cell-based regenerative medicine continues to show particular promise to jumpstart and facilitate tissue reconstruction, it has emergently been recognized that the regeneration process relies on coordinated behaviors of stem cells under tight functional control by niche factors [1–3]. Generally, local niche signals from extracellular matrix (ECM) microenvironments significantly affect stem cell migration, proliferation, lineage commitment, and arrangement [4–6]. Specifically, surrounding biophysical signals, such as soluble factors, membrane interactions, and matrix characteristics, converge to provide a multifaceted, diverse stem cell microenvironment for highly specific tissue morphogenesis and regeneration [7–9]. With the tissue engineering biotechnology advanced, various scaffold materials have been tailored and multiple cytokines have been finely adjusted to mimic the natural tissue ECM niche hoping to promote stem cell-based regeneration [10–12]. However, current manipulated niche lacks persistent bioactive beneficial stimulus as to the natural stem cell microenvironment, often leading to failure of

Bing-Dong Sui and Bin Zhu contributed equally to this work.

complete de novo regeneration. In this regard, topical favorable regenerative microenvironments can be efficiently established via the cell aggregate technique, which has been recognized as a promising concept for scaffold-free high-density stem cell delivery with adhesion molecules on the cell–surface and cell–cell interactions remaining intact [13–15]. Accordingly, this technology could preserve the self-produced, tissue-specific ECM thus mimicking natural stem cell niche in terms of complex mechanical, chemical, and biological properties, allowing encapsulated stem cells to rapidly acquire information of dynamic populational conditions via numerous signaling molecules, thus exhibiting homogenous or synergic regenerative behaviors as quorum sensing and responding. Meanwhile, engineered ECM in cell aggregates can provide local beneficial microenvironment for stem cell differentiation, the immense regenerative potential of which has been proved in a variety of hard and soft tissue regeneration [16–18]. Furthermore, this stem cell-based self-assembly approach to recreate the regenerative niche can be optimized using selected small molecule compounds or multiple layers to enhance tissue reconstruction performance against diseased recipient microenvironments [19–21].

In this protocol, we describe the basic methodology of establishing stem cell aggregates using bone marrow mesenchymal stem cells (BMMSCs), one of the most popular candidates for current regenerative application [1, 22], as the model as well as the quality check indexes prior to in vivo application.

2 Materials

2.1 Reagents

1. Basal cell culture media [23, 24]: Prepare alpha-minimum essential medium (α-MEM, Gibco #1864075) with 5%, 10%, or 20% fetal bovine serum (FBS, Gembio #900-208), 2 mM L-glutamine (Gibco #25030081), and 100 U/ml penicillin-–streptomycin (Gibco #15140122). Filter sterilize and store at 4 °C
2. Phosphate-buffered saline (PBS): pH 7.4 (Gibco #10010023), sterilize, and store at 4 °C
3. 0.25% Trypsin–1 mM ethylenediaminetetraacetic acid (EDTA) for cell digestion (Gibco #25200056): sterilize and store at 4 °C
4. Trypan blue solution: 0.4% (Lonza #17-942E)
5. Cell aggregate-inducing media [6, 16]: basal media (5% FBS) containing 50 μg/ml vitamin C (ascorbic acid, Sigma-Aldrich #A4403), and store at 4 °C away from light. Vitamin C stocking is prepared in distilled water, filter-sterilized, and stored in −20 °C away from light

6. Optional scaffold materials: β-tricalcium phosphate bioceramic (β-TCP, Sigma-Aldrich #Z687529) or other scaffolds can be used for certain regenerative purposes [6, 16]
7. Fixative 4% paraformaldehyde: 40 g paraformaldehyde (Sigma-Aldrich #P6148) dissolved in 1000 ml PBS with pH value adjusted to 7.4
8. An electrolyte containing 0.5% hydrofluoric acid (Sigma-Aldrich #339261) and 1 M phosphoric acid (Sigma-Aldrich #466123)
9. Dehydration: 70, 80, 90, and 100% ethanol (Azer Scientific #ES631) prepared by mixing with distilled water; xylene (Azer Scientific #ES609)
10. Embedding: Paraplast Plus (McCormick Scientific #39502004)
11. Hematoxylin and eosin (H&E) stain kit (Vector Laboratories #H-3502)
12. Trichrome stain (Masson) kit (Sigma-Aldrich #HT15)
13. Antigen retrieval: 10 mM sodium citrate dissolved in distilled water, pH 6.0 (Calbiochem #567446)
14. 3% Hydrogen peroxide solution: prepared with 30% hydrogen peroxide (Fisher Scientific #S25360) and distilled water, stored at 4 °C
15. 5% Bovine serum albumin (BSA): 5 g BSA (Gembio #700-100P) dissolved in 100 ml PBS and stored at 4 °C
16. Primary antibodies: a rabbit anti-mouse Collagen I antibody (Abcam #ab34710), a rabbit anti-mouse Periostin antibody (Abcam #ab14041), a rabbit anti-mouse Fibronectin antibody (Abcam #ab2413), and a rabbit anti-mouse Integrin β1 antibody (Abcam #ab149471). Prepare at a concentration of 1:100 in PBS just before use
17. A horse radish peroxidase (HRP)-conjugated goat anti-rabbit secondary antibody (Abcam #ab7090) at a concentration of 1:200 in PBS, prepared just before use
18. An HRP-based detection system (Dako #K5007)
19. Hematoxylin for counterstaining (Sigma-Aldrich #GHS316)
20. Permount mounting medium (VWR International #17986-01)

2.2 Appliances

1. Biological safety cabinet for cell culture
2. Sterilized surgical forceps and scissors for mice
3. Vacuum-driven filter system (Genesee Scientific #25-227)
4. Syringe (BD Biosciences #302995) and 25G × 7/8 needles (BD Biosciences #305124)
5. 70-μm Cell strainers (VWR International #10199-656)

6. 10-cm Culture dishes (Genesee Scientific #32-103) and 12-well culture plates (Genesee Scientific #25-106)
7. Centrifuge (Beckman Coulter #Allegra™ X-22R)
8. Hemocytometer
9. Pipettes, sterilized tips, and tubes
10. Scanning electron microscopy (SEM) (Hitachi #S-4800) and light microscopy (Olympus #BX-51)
11. Paraffin-embedding station
12. Embedding boxes
13. Rotary microtome (Thermo Scientific #HM325)
14. Tissue floating bath (Premiere #XH-1001)
15. Slides warmer (Premiere #XH-2001)
16. SuperFrost Plus glass slides (VWR International #48311-703)
17. Slip-Rite cover glass (Thermo Scientific #102450)
18. Tissue cassettes
19. Glass Coplin jars
20. Mini PAP Pen (Invitrogen #00-8877)
21. Moisture chamber for immunostaining
22. Refrigerator and oven

3 Methods

3.1 Cell Isolation and Culture

1. Sacrifice mice (C57/Bl6, 8-week old) by cervical dislocation and rinse the animal skin in 70% ethanol. Make incisions near the hind limbs and dissect the femora and tibia with surrounding muscle and connective tissues but without any skin (*see* **Note 1**). Store the limbs on ice in cold PBS.
2. Perform further dissection in the biological safety cabinet. Remove the muscle and connective tissues from the skeleton in cold PBS (*see* **Note 2**). Cut the ends of the tibia and femora just below the epiphysis to expose the marrow cavity. Store the skeleton in basal culture media with 20% FBS.
3. Instantly insert the needle attached to the syringe containing 10-ml basal culture media with 20% FBS into the bone marrow space. Flush the marrow plug out of the cut end of the bone till the whole marrow space turns white (*see* **Note 3**). Filter the bone marrow cell suspension through the strainer to remove any bone spicules or cell clumps.
4. Seeding cells in culture dishes (usually 2 dishes for 1 mouse) in a total 10-ml basal culture media with 20% FBS (*see* **Note 4**).

Incubate the dishes at 37 °C with 5% CO_2 in a humidified chamber without disturbing them.

5. After overnight incubation, remove the nonadherent cells by repeated PBS washing and add fresh 10-ml basal culture media with 20% FBS (*see* **Note 5**). Change the culture media every 3 days. After about 10 days, cell colonies varying in size can be observed reaching 60–70% confluence (Fig. 1a).
6. To passage, wash cells with PBS and incubate cells in 4-ml 0.25% trypsin–1 mM EDTA at 37 °C for 5 min. Neutralize trypsin by adding 4-ml basal culture media with 20% FBS. Gently collect the lifted cells by using pipettes (*see* **Note 6**). Centrifuge the digested cells in 800 × *g* for 5 min.
7. Seed cells in fresh dishes (2 dishes at passage 0 for only 1 dish at passage 1) in 10-ml basal culture media with 20% FBS. Change the culture media every 3 days, BMMSCs at the 1st passage and later grow in a dispersed pattern (Fig. 1b).
8. Digest and collect cells as stated above when reaching confluence (typically about 6 days in passage 1). Determine the yield and viability of cells by Trypan blue exclusion and counting on a hemocytometer (about at least 2×10^6 cells can be harvested from 1 dish of 1st-passaged BMMSCs).
9. Seed cells in 12-well culture plates at 2×10^5 cells per well in 1.5 ml basal culture media with 10% FBS. After 2–3 days, when reaching 95% confluence, the cells will be ready for cell aggregate construction (*see* **Note 7**).
10. For BMMSC identification, the 2nd-passaged cells can be used for flow cytometric analysis of surface markers and multiple lineage differentiation [25, 26].

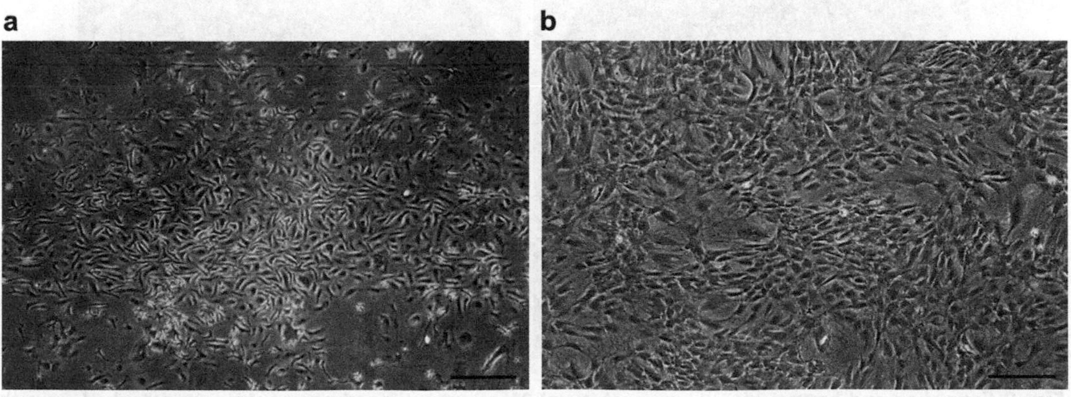

Fig. 1 Culture of mouse BMMSCs. (**a**) Colony formation of primary BMMSCs under microscopy. (**b**) BMMSCs at the 1st passage grow in a dispersed pattern under microscopy. Bars: 250 μm

3.2 Construction and Optimization of Cell Aggregates

1. Prepare 2nd-passaged BMMSCs in 12-well culture plates as stated above. When reaching 95% confluence, change the culture media to the cell aggregate-inducing media containing 5% FBS and 50 μg/ml vitamin C. Direct light exposure should be avoided hereafter (*see* **Note 8**).
2. The cell aggregate-inducing media are changed every 2 days. Cells are observed under light microscopy at each time to monitor the induction process.
3. After about 10–14 days (according to different stem cell types, often about 12 days for BMMSCs), curling edge can be seen under light microscopy with membrane-shaped aggregates. In some wells, the cell aggregates are detachable to the bottom of the well (*see* **Note 9**).
4. Gently aspirate most culture media from the well, leaving only enough media on surface to keep moisture of the cell aggregates. Carefully use sterilized forceps to detach the cell aggregates from the bottom of the well, in an order evenly from peripheral part to the central part (Fig. 2a) (*see* **Note 10**).
5. If needed, scaffold materials such as β-TCP for bone regeneration can be put in the geometric center of the cell aggregates and then be wrapped (*see* **Note 11**). However, the cell aggregates surely can be used in a scaffold-free manner.
6. For multiple-layered cell aggregates, the wrapped aggregates of the first layer can be put in the geometric center of the next layer to be wrapped again. As much as three layers of cell aggregates can be obtained in this method (*see* **Note 12**).

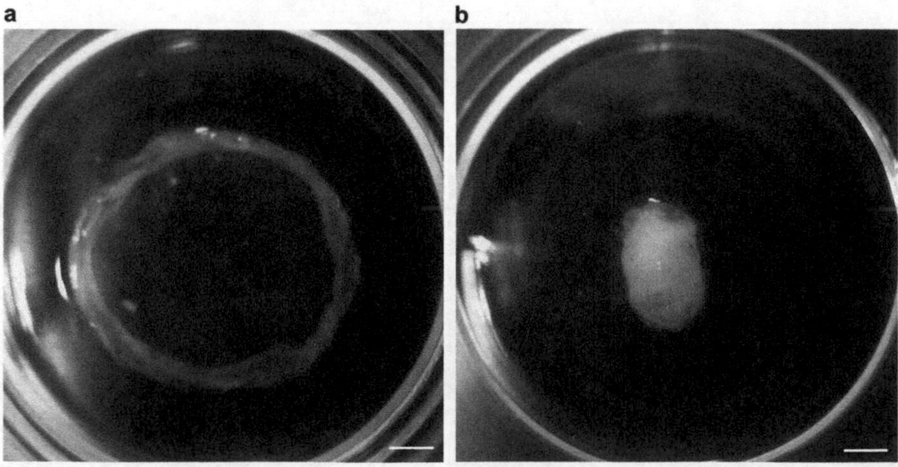

Fig. 2 Formation and harvest of cell aggregates. (**a**) Membrane-shaped aggregates are detached from the bottom of the well, in an order evenly from peripheral part to the central part. (**b**) Wrapped cell aggregates after harvest. Bars: 2.5 mm

Also, aggregates derived from different stem cell populations can be applied.

7. Keep culturing the wrapped cell aggregates (Fig. 2b) in enough inducing media for 24 h, and the aggregates will be ready for implantation or quality check as stated below.

3.3 Morphological Observation of Cell Aggregates

1. As the cell aggregates form, healthy aggregates possess a white semitransparent membrane shape, with soft but tough characteristics that can be wrapped by mechanical force without being damaged (*see* **Note 13**). Under microscopy, cell aggregates with compact woven pattern can be observed (Fig. 3a).
2. For histological quality check, cell aggregates are fixed in 4% paraformaldehyde overnight at 4 °C, washed with PBS, and dehydrated sequentially through 70, 80, 90, and 100% ethanol (each for 10 min twice) and xylene (for 30 min followed by 15 min).

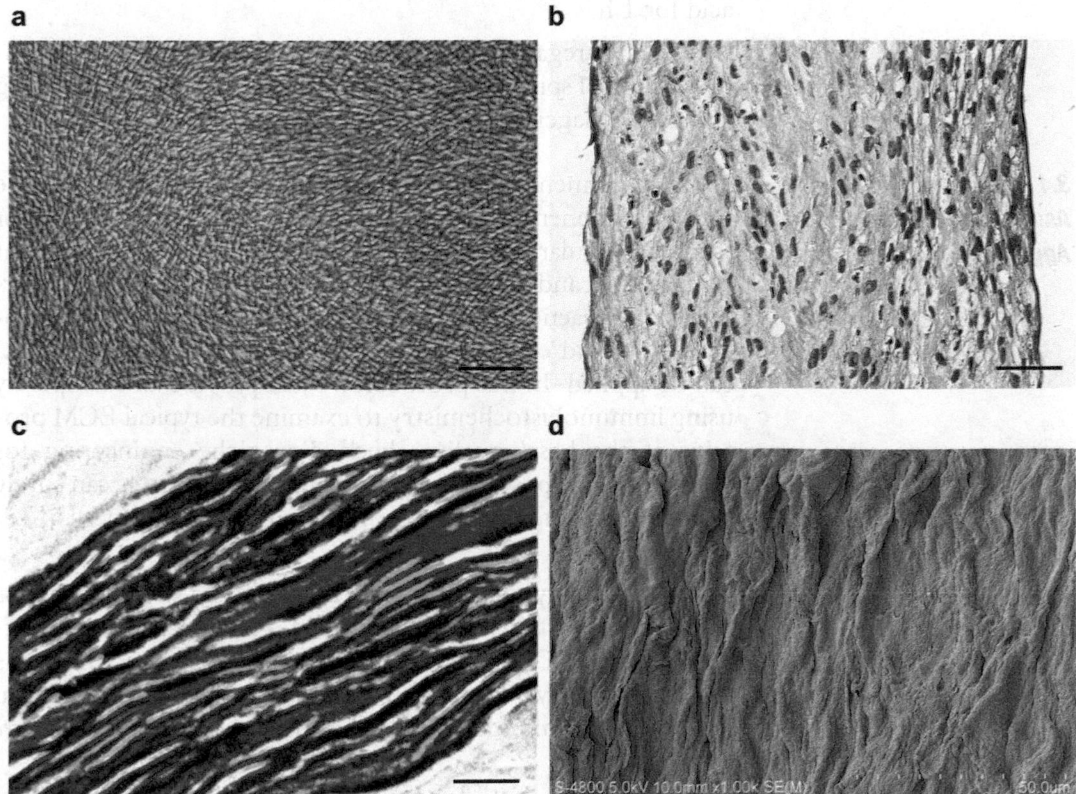

Fig. 3 Morphological observation of cell aggregates. (**a**) Good-quality cell aggregates with compact woven pattern under microscopy. Bar: 250 μm. (**b**) H&E staining demonstrates distribution of cells surrounded by continuous ECM in cell aggregates. Bar: 100 μm. (**c**) Masson's Trichrome staining illustrates regularly interlaced ECM. Bar: 100 μm. (**d**) SEM observation of spindle-shaped BMMSCs encapsulated by strong compact collagen fibers. Conditions of SEM and the bar are indicated

3. Embed the cell aggregates in paraffin. 5-μm Thickness longitudinally sectioned slices are prepared.
4. Dewaxing sequentially through xylene (for 3 min twice) and 100, 90, 80, and 70% ethanol (each for 1 min twice) to water.
5. Stain with H&E according to manufacturer's instructions of the commercial kit as stated above. Cell and ECM arrangements of the aggregates can be observed under light microscopy (Fig. 3b). Good-quality aggregates demonstrate uninterrupted compact distribution of cells with at least one layer, surrounded by continuous ECM (*see* **Note 14**).
6. Slides are also stained with Masson's Trichrome kit according to manufacturer's instructions. Collagen deposition can be analyzed with the blue-stained area (Fig. 3c). Good-quality aggregates exhibit homogeneous, regularly interlaced ECM.
7. For ultrastructural quality check, cell aggregates are fixed in 4% paraformaldehyde overnight at 4 °C and anodized in an electrolyte containing 0.5% hydrofluoric acid and 1 M phosphoric acid for 1 h.
8. Observe aggregates under SEM. Good-quality aggregates display plenty of spindle-shaped BMMSCs encapsulated by strong compact collagen fibers (*see* **Note 14**) (Fig. 3d).

3.4 Niche Assessment of Cell Aggregates

1. Niche assessment of cell aggregates is complex, which include ECM components (such as Fibronectin) and secretory factor (such as vascular endothelial growth factor, VEGF) analysis at both mRNA and protein expression levels. Generally, polymerase chain reaction (PCR), western blotting, immunohistochemistry, and enzyme-linked immunosorbent assay (ELISA) can be applied. In this protocol, we just provide an example by using immunohistochemistry to examine the typical ECM proteins as the basal quality check. For niche engineering for determined purposes, other factors in the aggregates can surely be investigated following the published methodologies [6, 13, 27].
2. For good-quality aggregates, the ECM should strongly express proteins such as Collagen I, Periostin, Fibronectin, and Integrin β1. These ECM molecules not only provide structural basis for the integrity of aggregates but also serve as the crucial factors for communication and adhesion of stem cells (*see* **Note 14**).
3. For immunohistochemistry, cell aggregates are fixed in 4% paraformaldehyde overnight at 4 °C, washed with PBS, and dehydrated sequentially through 70, 80, 90, and 100% ethanol (each for 10 min twice) and xylene (for 30 min followed by 15 min).

4. Embed the cell aggregates in paraffin. 5-μm Thickness longitudinally sectioned slices are prepared.
5. Dewaxing sequentially through xylene (for 3 min twice) and 100, 90, 80, and 70% ethanol (each for 1 min twice) to water.
6. Antigen retrieval: Bring slides to a boiled 10-mM sodium citrate buffer, and then maintain at a sub-boiling temperature for 10 min. Cool slides on benchtop for 30 min.
7. Draw a circle around the specimens using the Mini PAP Pen, and treat the samples by 3% hydrogen peroxide solution. Wash with PBS for three times, 5 min each.
8. Block specimens with 5% BSA for 1 h at room temperature. Prepare primary antibodies for Collagen I, Periostin, Fibronectin, and Integrin β1 as indicated, aspirate blocking solution, and apply diluted primary antibodies. Incubate slides overnight at 4 °C.
9. Wash with PBS for three times, 5 min each. Prepare the HRP-conjugated secondary antibody as indicated, apply the diluted secondary antibody, and incubate specimens for 1 h at room temperature.
10. Wash with PBS for three times, 5 min each. Detect the signals according to the manufacturer's instructions of the HRP-based detection system kit.
11. Counterstain with hematoxylin for 1 min at room temperature. Wash with distilled water, treat sequentially through 95% and 100% ethanol and xylene (each for 1 min twice), and mounting.
12. Observe under a light microscopy (Fig. 4a–d).

4 Notes

1. To avoid contamination from mouse skin, make incisions closely to the ankle and carefully pull the full layer of skin inside-out reversely to the waist.
2. Be careful not to break the skeleton in clearing soft tissues. Keep low temperature to preserve stem cell activity for culture.
3. If any red marrow tissues remain in the marrow space, put the needle directly onto them to flush. Solid marrow contents can be further dispersed to suspensions by using either scissors or pipettes.
4. Use pipettes to seed cells as evenly as possible in the culture dish. This step is critical to obtain the many colonies with similar sizes thus elevating the harvest.

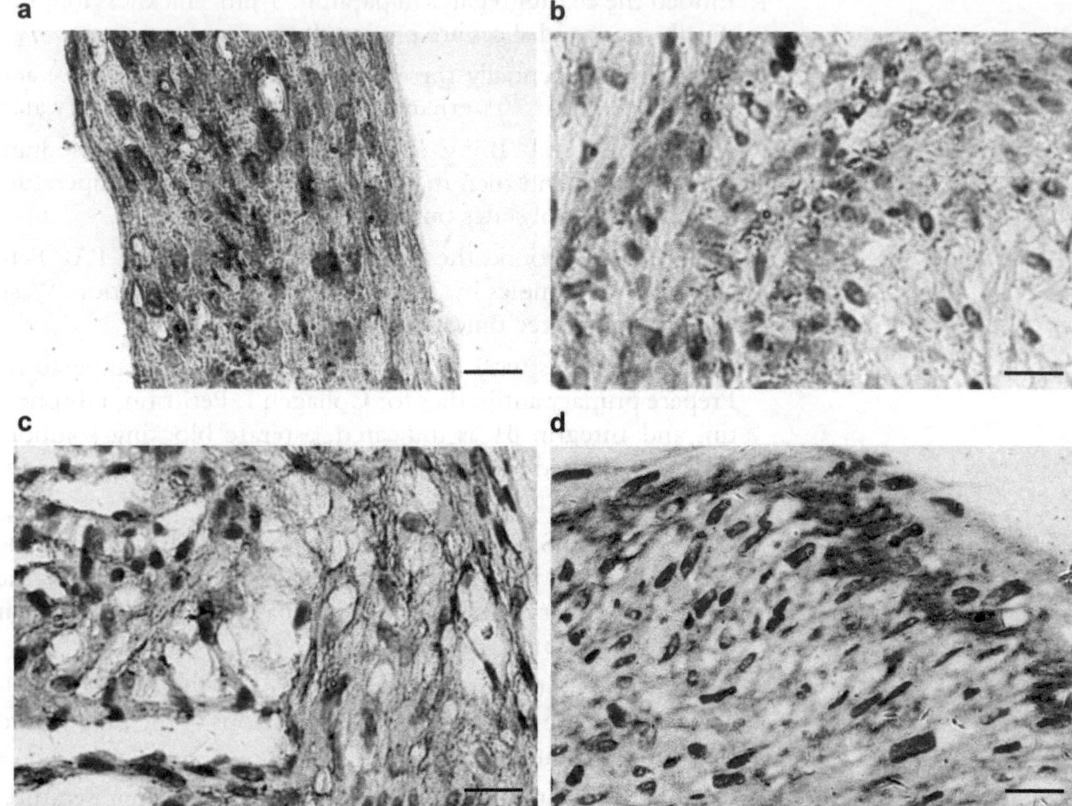

Fig. 4 ECM niche assessment of cell aggregates. (**a**) Immunohistochemistry evaluation of Collagen I in cell aggregates. (**b**) Immunohistochemistry evaluation of Integrin β1 in cell aggregates. (**c**) Immunohistochemistry evaluation of Fibronectin in cell aggregates. (**d**) Immunohistochemistry evaluation of Periostin in cell aggregates. Bars: 50 μm

5. BMMSCs should have adhered tightly onto the culture dish surface after overnight incubation, though may not be observable. Wash the dish with PBS till no suspending cells remain. Rather pure primary BMMSCs can be acquired by applying this simple technique.
6. Primary BMMSC colonies are not easy to be digested. Enough trypsin incubation time at 37 °C followed by pipetting is needed. However, over-incubation with trypsin will surely induce cell damages with unwanted cell types (mostly macrophages) detached off the culture dish surface. An ideal digestion efficacy should be controlled within 80–90%.
7. Stem cells for aggregate induction should be kept with good viability with less than three passages and enough density. Particularly, original cell density with a near-full confluence is necessary for successful aggregate induction. Virtually, no space should be observed left among the cells before beginning of the induction. Do not change to the aggregate-inducing

media too early. Otherwise, more induction time will be needed for aggregate formation with increased risks to get thin, fragile aggregates.

8. Lower the serum concentration to 5% during cell aggregate induction. This method can slow down cell proliferation to coordinate ECM secretion and apposition. Also, do not frequently move the plates during aggregate induction.
9. Timing of aggregate harvest should be controlled well. While slight edge curls are the sign of maturity, over-curling causing partial overlapped cell layers should be avoided.
10. Gradually wrap the cell aggregates with patience. The whole harvesting process takes about 5 min.
11. Keep moisture of the cell aggregates. Nevertheless, too much media left is no good for gluing scaffolds with the aggregates.
12. Too many layers to form cell aggregates will result in necrosis of cells in the central part due to lack of enough nutrition or ventilation.
13. Bad-quality cell aggregates, on the other hand, will easily be broken by slight mechanical force to show rough edges or cracks, because of little ECM deposition by stem cells at low viability.
14. Usually, quality check of cell aggregates can be achieved simply during the harvest process. Bad-quality cell aggregates indicate unsatisfied reconstruction of regenerative niche and will fail in regenerative applications. The engineered stem cell niche can be further confirmed via histological methods.

Acknowledgments

This work was supported by grants from the National Key Research and Development Program of China (2016YFC1102900 and 2016YFC1101400), the General Program of National Natural Science Foundation of China (81570937 and 81470710), and the State Scholarship Fund of China (201603170205).

References

1. Sui BD, Hu CH, Liu AQ, Zheng CX, Xuan K, Jin Y (2017) Stem cell-based bone regeneration in diseased microenvironments: challenges and solutions. Biomaterials. https://doi.org/10.1016/j.biomaterials.2017.10.046
2. Paschos NK, Brown WE, Eswaramoorthy R, Hu JC, Athanasiou KA (2015) Advances in tissue engineering through stem cell-based co-culture. J Tissue Eng Regen Med 9 (5):488–503. https://doi.org/10.1002/term.1870
3. Sui BD, Hu CH, Zheng CX, Shuai Y, He XN, Gao PP, Zhao P, Li M, Zhang XY, He T, Xuan K, Jin Y (2017) Recipient glycemic micro-environments govern therapeutic effects of mesenchymal stem cell infusion on osteopenia. Theranostics 7(5):1225–1244. https://doi.org/10.7150/thno.18181

4. Reilly GC, Engler AJ (2010) Intrinsic extracellular matrix properties regulate stem cell differentiation. J Biomech 43(1):55–62. https://doi.org/10.1016/j.jbiomech.2009.09.009
5. Engler AJ, Sen S, Sweeney HL, Discher DE (2006) Matrix elasticity directs stem cell lineage specification. Cell 126(4):677–689. https://doi.org/10.1016/j.cell.2006.06.044
6. Zhu B, Liu W, Zhang H, Zhao X, Duan Y, Li D, Jin Y (2017) Tissue-specific composite cell aggregates drive periodontium tissue regeneration by reconstructing a regenerative microenvironment. J Tissue Eng Regen Med 11(6):1792–1805. https://doi.org/10.1002/term.2077
7. Discher DE, Janmey P, Wang YL (2005) Tissue cells feel and respond to the stiffness of their substrate. Science 310(5751):1139–1143. https://doi.org/10.1126/science.1116995
8. Huebsch N, Arany PR, Mao AS, Shvartsman D, Ali OA, Bencherif SA, Rivera-Feliciano J, Mooney DJ (2010) Harnessing traction-mediated manipulation of the cell/matrix interface to control stem-cell fate. Nat Mater 9(6):518–526. https://doi.org/10.1038/nmat2732
9. Sui BD, Hu CH, Zheng CX, Jin Y (2016) Microenvironmental views on mesenchymal stem cell differentiation in aging. J Dent Res 95(12):1333–1340. https://doi.org/10.1177/0022034516653589
10. Atala A, Kasper FK, Mikos AG (2012) Engineering complex tissues. Sci Transl Med 4 (160):160rv112. https://doi.org/10.1126/scitranslmed.3004890
11. Chen G, Chen J, Yang B, Li L, Luo X, Zhang X, Feng L, Jiang Z, Yu M, Guo W, Tian W (2015) Combination of aligned PLGA/Gelatin electrospun sheets, native dental pulp extracellular matrix and treated dentin matrix as substrates for tooth root regeneration. Biomaterials 52:56–70. https://doi.org/10.1016/j.biomaterials.2015.02.011
12. Prewitz MC, Seib FP, von Bonin M, Friedrichs J, Stissel A, Niehage C, Muller K, Anastassiadis K, Waskow C, Hoflack B, Bornhauser M, Werner C (2013) Tightly anchored tissue-mimetic matrices as instructive stem cell microenvironments. Nat Methods 10 (8):788–794. https://doi.org/10.1038/nmeth.2523
13. An Y, Wei W, Jing H, Ming L, Liu S, Jin Y (2015) Bone marrow mesenchymal stem cell aggregate: an optimal cell therapy for full-layer cutaneous wound vascularization and regeneration. Sci Rep 5:17036. https://doi.org/10.1038/srep17036
14. Iwata T, Washio K, Yoshida T, Ishikawa I, Ando T, Yamato M, Okano T (2015) Cell sheet engineering and its application for periodontal regeneration. J Tissue Eng Regen Med 9(4):343–356. https://doi.org/10.1002/term.1785
15. Shang F, Ming L, Zhou Z, Yu Y, Sun J, Ding Y, Jin Y (2014) The effect of licochalcone A on cell-aggregates ECM secretion and osteogenic differentiation during bone formation in metaphyseal defects in ovariectomized rats. Biomaterials 35(9):2789–2797. https://doi.org/10.1016/j.biomaterials.2013.12.061
16. Shang F, Liu S, Ming L, Tian R, Jin F, Ding Y, Zhang Y, Zhang H, Deng Z, Jin Y (2017) Human umbilical cord MSCs as new cell sources for promoting periodontal regeneration in inflammatory periodontal defect. Theranostics 7(18):4370–4382. https://doi.org/10.7150/thno.19888
17. Liu Y, Ming L, Luo H, Liu W, Zhang Y, Liu H, Jin Y (2013) Integration of a calcined bovine bone and BMSC-sheet 3D scaffold and the promotion of bone regeneration in large defects. Biomaterials 34(38):9998–10006. https://doi.org/10.1016/j.biomaterials.2013.09.040
18. Dang PN, Solorio LD, Alsberg E (2014) Driving cartilage formation in high-density human adipose-derived stem cell aggregate and sheet constructs without exogenous growth factor delivery. Tissue Eng Part A 20 (23–24):3163–3175. https://doi.org/10.1089/ten.TEA.2012.0551
19. Sun J, Dong Z, Zhang Y, He X, Fei D, Jin F, Yuan L, Li B, Jin Y (2017) Osthole improves function of periodontitis periodontal ligament stem cells via epigenetic modification in cell sheets engineering. Sci Rep 7(1):5254. https://doi.org/10.1038/s41598-017-05762-7
20. Shuai Y, Liao L, Su X, Yu Y, Shao B, Jing H, Zhang X, Deng Z, Jin Y (2016) Melatonin treatment improves mesenchymal stem cells therapy by preserving stemness during long-term in vitro expansion. Theranostics 6 (11):1899–1917. https://doi.org/10.7150/thno.15412
21. Liu Y, Wang L, Kikuiri T, Akiyama K, Chen C, Xu X, Yang R, Chen W, Wang S, Shi S (2011) Mesenchymal stem cell-based tissue regeneration is governed by recipient T lymphocytes via IFN-gamma and TNF-alpha. Nat Med 17 (12):1594–1601. https://doi.org/10.1038/nm.2542
22. Sui B, Hu C, Jin Y (2016) Mitochondrial metabolic failure in telomere attrition-provoked

aging of bone marrow mesenchymal stem cells. Biogerontology 17(2):267–279. https://doi.org/10.1007/s10522-015-9609-5

23. Hu CH, Sui BD, Du FY, Shuai Y, Zheng CX, Zhao P, Yu XR, Jin Y (2017) miR-21 deficiency inhibits osteoclast function and prevents bone loss in mice. Sci Rep 7:43191. https://doi.org/10.1038/srep43191
24. Chen N, Sui BD, Hu CH, Cao J, Zheng CX, Hou R, Yang ZK, Zhao P, Chen Q, Yang QJ, Jin Y, Jin F (2016) microRNA-21 contributes to orthodontic tooth movement. J Dent Res 95(12):1425–1433. https://doi.org/10.1177/0022034516657043
25. Sui B, Hu C, Liao L, Chen Y, Zhang X, Fu X, Zheng C, Li M, Wu L, Zhao X, Jin Y (2016) Mesenchymal progenitors in osteopenias of diverse pathologies: differential characteristics in the common shift from osteoblastogenesis to adipogenesis. Sci Rep 6:30186. https://doi.org/10.1038/srep30186
26. Sui B, Hu C, Zhang X, Zhao P, He T, Zhou C, Qiu X, Chen N, Zhao X, Jin Y (2016) Allogeneic mesenchymal stem cell therapy promotes osteoblastogenesis and prevents glucocorticoid-induced osteoporosis. Stem Cells Transl Med 5(9):1238–1246. https://doi.org/10.5966/sctm.2015-0347
27. Zhao P, Sui BD, Liu N, Lv YJ, Zheng CX, Lu YB, Huang WT, Zhou CH, Chen J, Pang DL, Fei DD, Xuan K, Hu CH, Jin Y (2017) Anti-aging pharmacology in cutaneous wound healing: effects of metformin, resveratrol, and rapamycin by local application. Aging Cell 16(5):1083–1093. https://doi.org/10.1111/acel.12635

Methods in Molecular Biology (2019) 2002: 101–119
DOI 10.1007/7651_2018_181

Published online: 27 October 2018

Three-Dimensional Co-culture of Human Hematopoietic Stem/Progenitor Cells and Mesenchymal Stem/Stromal Cells in a Biomimetic Hematopoietic Niche Microenvironment

Marta H. G. Costa, Tiago S. Monteiro, Susana Cardoso, Joaquim M. S. Cabral, Frederico Castelo Ferreira, and Cláudia L. da Silva

Abstract

The development of cellular therapies to treat hematological malignancies has motivated researchers to investigate ex vivo culture systems capable of expanding the number of hematopoietic stem/progenitor cells (HSPC) before transplantation. The strategies exploited to achieve relevant cell numbers have relied on culture systems that lack biomimetic niche cues thought to be essential to promote HSPC maintenance and proliferation. Although stromal cells adhered to 2-D surfaces can be used to support the expansion of HSPC ex vivo, culture systems aiming to incorporate cell–cell interactions in a more intricate 3-D environment can better contribute to recapitulate the bone marrow (BM) hematopoietic niche in vitro.

Herein, we describe the development of a 3-D co-culture system of human umbilical cord blood (UCB)-derived $CD34^+$ cells and BM mesenchymal stem/stromal cell (MSC) spheroids in a microwell-based platform that allows to attain large numbers of spheroids with uniform sizes. Further comparison with a traditional 2-D co-culture system exploiting the supportive features of feeder layers of MSC is provided, while functional in vitro assays to assess the features of HSPC expanded in the 2-D vs. 3-D MSC co-culture systems are suggested.

Keywords Co-culture, Hematopoietic niche, Hematopoietic stem/progenitor cells (HSPC), Mesenchymal stem/stromal cells (MSC), Microwells, Spheroids

1 Introduction

Hematopoietic stem cells (HSC) have the potential to give rise to the entire hematopoietic lineage, originating both lymphoid and myeloid lineages, while retaining their self-renewal potential. Although hematopoietic stem/progenitor cells (HSPC) can be isolated from different sources (bone marrow (BM), umbilical cord blood (UCB), and mobilized peripheral blood), UCB-derived HSPC present less stringent human leukocyte antigen (HLA) matching, are less likely to be contaminated with blood-borne viruses, and are relatively easy to collect. Particularly, the expression of CD34 antigen can be exploited to enrich UCB cells in stem/progenitor cells for transplantation, as the CD34$^+$ cell population is thought to comprise the more primitive hematopoietic cells.

Due to the limited numbers of hematopoietic cells that can be isolated from an UCB unit, their expansion in vitro is particularly relevant to widen their therapeutic applications. Nonetheless, whereas the hematopoietic niche allows HSC to self-renew and maintain throughout an entire individual's lifetime, culture of HSPC in vitro is often accompanied by their differentiation.

The development of cell culture platforms capable of mimicking a hematopoietic niche-like microenvironment could contribute to retain the functionality of HSPC cultured in vitro. Indeed, the concept of a HSC niche has been first proposed by Schofield in the late 1970s [1] and imaging studies depicting the localization of HSC within the BM niche suggest that their association with niche cells is key to retain a quiescent state [2].

BM mesenchymal stem/stromal cells (BM MSC) are an important component of the hematopoietic niche, supporting hematopoiesis and regulating the maintenance of primitive HSC [2, 3]. Although the mechanisms provided by MSC involved on governing the fate of HSC are not fully understood, secretion of soluble cues (e.g., stem cell factor (SCF), thrombopoietin (TPO), and FMS-like tyrosine kinase-3 ligand (Flt-3L)) is thought to support self-renewal of HSC [4, 5], while expression of cell adhesion molecules (e.g., N-cadherin [6, 7]) and extracellular matrix (ECM) components (e.g., osteopontin and tenascin [8, 9]) are also likely to impact cell function [10]. Furthermore, the HSPC behavior is not only modulated by a specific cocktail of cytokines and ECM components but also by a 3-D architecture provided by the cellular environment present in the BM.

The importance of a 3-D HSPC/MSC co-culture microenvironment to modulate the activity of HSPC was highlighted by Jing and colleagues [10], who showed the existence of a spatial relationship between HSPC and MSC during ex vivo expansion, and further exploited by Cook and co-workers [11], who developed a 3-D HSPC/MSC co-culture platform to support the expansion of more primitive hematopoietic cells. Posterior studies have hypothesized that culture of MSC as 3-D spheroids could contribute to mimic stem cell niche interactions, regulating HSC maintenance, self-renewal, or differentiation in vitro [12–15].

3-D cellular interactions established in MSC spheroids have shown to promote changes on their biological activity, particularly regarding their ability to support stem cell properties [14]. 3-D MSC spheroids express higher levels of niche-specific ECM components [15], and of Prx1 and Nestin, markers of mesenchymal niche cells that support HSC in the BM [2, 14, 16]. Moreover, CD34$^+$ cells co-cultured with 3-D MSC spheroids generated higher levels of the more primitive CD34$^+$CD90$^+$ subpopulation when compared to 2-D co-cultured hematopoietic cells [14], suggesting that 3-D MSC constitute an activated niche to favor HSPC self-renewal. On the other hand, the differences on the paracrine activity of 2-D or 3-D cultured MSC could potentially impact their ability to regulate the lineage commitment of hematopoietic cells and alter their cell cycle state.

Although several techniques have been explored to form spheroids (spinner flasks [17], hanging drops [18], culture in low attachment surfaces [19], and magnetic cell levitation [20]), the resulting spheroids usually present heterogeneous sizes and require multiple iterations to achieve the required number of aggregates. Alternatively, microwell arrays constitute an attractive tool to generate spheroids with uniform dimensions in a high-throughput manner [11, 12]. Importantly, co-culture of HSPC and MSC in a multiwell plate format can mimic hematopoietic niche-like features by maximizing cell–cell interactions and potentiating the paracrine and autocrine signals that are rapidly diluted in the large volumes of medium present in standard culture systems. Different cell seeding densities and microwell diameters can help controlling the size of the spheroids while allowing cells to be evenly seeded into each individual microwell. Importantly, secreted cytokines and multicellular interactions established within the HSPC/MSC spheroid could be facilitated by their retention in microwells.

Herein, we describe a protocol to culture HSPC in conventional 2-D co-cultures supported by feeder layers of MSC or in a biomimetic hematopoietic niche-like environment, either supported by 3-D MSC spheroids freely dispersed in the culture plates or kept in microwell arrays (Fig. 1). The functional assays suggested to assess the features of HSPC expanded in the 2-D vs. 3-D MSC co-culture systems include: identification of colony-forming units (CFUs) [21, 22] and cobblestone area-forming cells (CAFCs) [21] (two assays frequently used to quantify hematopoietic progenitor cells based on their clonal ability to originate multiple hematopoietic colonies—multipotential and lineage-restricted progenitors—or to form cobblestone areas underneath stromal cells—indicative of the potential repopulating ability of the hematopoietic cell, respectively) and cell adhesion and migration [23] assays, key to assess the capacity of co-cultured HSPC to home and migrate towards the hematopoietic niche.

2 Materials

2.1 Thawing and Expansion of Human Bone Marrow (BM)-Derived MSC

1. Laminar flow hood
2. Cell culture incubator with CO_2, temperature, and humidity control
3. Centrifuge
4. Inverted microscope equipped with ultraviolet (UV) light

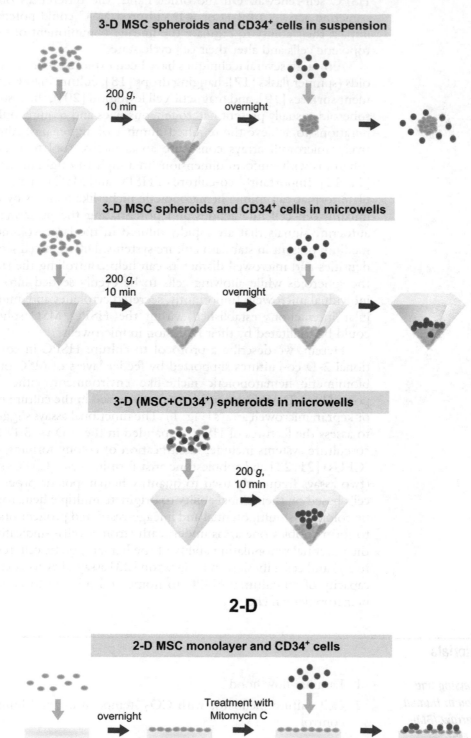

Fig. 1 Schematic representation of 3-D and 2-D co-culture systems of $CD34^+$-enriched cells and MSC

5. Water bath set to 37 °C
6. Hemocytometer
7. T-flasks T75 (BD Biosciences)
8. Polypropylene conical tubes (15/50 mL, BD Biosciences)
9. Thawing medium: Dulbecco's modified Eagle's medium (DMEM), 20% fetal bovine serum (FBS) (Life Technologies), and 1% antibiotic/antimycotic (A/A). Store at 4 °C
10. MSC culture medium: low-glucose DMEM supplemented with 10% FBS MSC-qualified (Life Technologies) and 1% A/A (Life Technologies). Store at 4 °C
11. Phosphate-buffered saline (PBS) solution. Dissolve PBS powder (Life Technologies) in 1 L of distilled water. Filter the solution using a 0.22-μm filter and store at room temperature (RT)
12. 0.05% (w/v) Trypsin (Life Technologies)—1 mM ethylenediaminetetraacetic acid solution (EDTA) (Sigma). Store at 4 °C
13. Trypan blue stain 0.4% (Life Technologies). Store at RT

2.2 Thawing and Expansion of MS-5 Murine Stromal Cells

1. Laminar flow hood
2. Cell culture incubator with CO_2, temperature, and humidity control
3. Centrifuge
4. Inverted microscope equipped with UV light
5. Water bath set to 37 °C
6. Hemocytometer
7. T-flasks T75 (BD Biosciences)
8. Polypropylene conical tubes (15/50 mL, BD Biosciences)
9. Thawing medium: DMEM 20% FBS 1% A/A (Life Technologies). Store at 4 °C
10. MS-5 culture medium: low-glucose DMEM supplemented with 10% FBS (Life Technologies) and 1% A/A (Life Technologies). Store at 4 °C
11. PBS solution. Dissolve PBS powder (Life Technologies) in 1 L of distilled water. Filter the solution using a 0.22-μm filter and store at RT
12. 0.05% (w/v) Trypsin (Life Technologies)—1 mM EDTA (Sigma). Store at 4 °C
13. Trypan blue stain 0.4% (Life Technologies). Store at RT

2.3 Preparation of Microwells

1. Centrifuge
2. Inverted microscope

3. Falcon® 24-well plates
4. 250 mL Glass bottle
5. Autoclave
6. Microwave
7. Polydimethylsiloxane (PDMS) molds containing 400 μm × 400 μm-sized etched pyramidal profiles confined to 15.6 mm diameter circumferences
8. Tweezers (autoclaved at 121 °C, 20 min)
9. Agarose (SeaKem® LE Agarose) 3% (w/v). Store at RT
10. Deionized water
11. DMEM 1% A/A. Store at 4 °C
12. Culture medium: low-glucose DMEM 10% FBS 1% A/A. Store at 4 °C

2.4 Formation of Spheroids

1. Cell culture incubator with CO_2, temperature, and humidity control
2. Centrifuge
3. Inverted microscope
4. 24-Well Ultralow Attachment plates (Corning)
5. Culture medium: low-glucose DMEM 10% FBS 1% A/A. Store at 4 °C

2.5 Preparation of a Feeder Layer of BM MSC

1. Cell culture incubator with CO_2, temperature, and humidity control
2. Falcon® 24-well plates
3. Mitomycin C (Sigma) (0.5 μg/mL prepared in DMEM 10% FBS)
4. Culture medium: DMEM 10% FBS 1% A/A. Store at 4 °C

2.6 Isolation and Cryopreservation of UCB Mononuclear Cells (MNC)

1. Blood bag
2. Syringes, 50 mL
3. 18G needle
4. Bottle
5. Pasteur pipette
6. Hemocytometer
7. Centrifuge
8. Polypropylene conical tubes (15/50 mL, BD Biosciences)
9. Cryogenic vials
10. Liquid/vapor phase nitrogen tank
11. Freezer at −80 °C

12. Recovery™ Cell Culture Freezing Medium (Thermo Fisher)
13. Ficoll-Paque medium (1.077 g/mL; GE Healthcare)
14. PBS
15. PBS supplemented with 2 mM EDTA (Gibco) solution. Store at 4 °C
16. Türk's reagent (Millipore). Store at RT

2.7 CD34⁺-Enrichment of UCB HSPC

1. Water bath set to 37 °C
2. Centrifuge
3. Thawing medium: DMEM 20% FBS containing DNAse I (5 ng/mL) (Roche Applied Science)
4. Trypan blue stain 0.4% (Life Technologies). Store at RT
5. Magnetic-activated cell sorting (MACS) buffer: PBS supplemented with 2 mM EDTA (Gibco) and 0.5% BSA (Sigma). Store at 4 °C
6. CD34 MicroBead kit (Miltenyi)
7. MACS columns (Miltenyi)
8. MACS separator (Miltenyi)
9. Pre-separation filters with a 30-μm nylon mesh
10. Polypropylene conical tubes (15/50 mL, BD Biosciences)
11. Trypan blue solution 0.4%. Store at RT
12. Culture medium: QBSF-60 serum-free medium (Quality Biological, Inc.) supplemented with 1% A/A and stem cell factor (SCF) (60 ng/mL), FMS-like tyrosine kinase-3 ligand (Flt-3L) (55 ng/mL), basic fibroblast growth factor (bFGF) (5 ng/mL), and thrombopoietin (TPO) (50 ng/mL) (all from PeproTech®; optimization of the cytokine cocktail is described in [21]). Store at 4 °C

2.8 Co-culture of Human UCB CD34⁺-Enriched Cells and BM MSC

1. Cell culture incubator with CO_2, temperature, and humidity control
2. Falcon® 24-well plates
3. Ultralow Attachment 24-well plates (Corning® Costar®)
4. Polypropylene conical tubes (15/50 mL, BD Biosciences)
5. Reversible cell strainer (37 μm, STEMCELL Technologies)
6. Culture medium: QBSF-60 serum-free medium (Quality Biological, Inc.) supplemented with 1% A/A and SCF (60 ng/mL), Flt-3L (55 ng/mL), bFGF (5 ng/mL), and TPO (50 ng/mL) (all from PeproTech®). Store at 4 °C

2.9 Immunophenotypic and Functional Characterization of UCB CD34+-Enriched Cells Co-cultured with BM MSC

2.9.1 Cell Counting and Viability

1. Inverted microscope
2. PBS. Store at RT
3. Polypropylene conical tubes (15/50 mL, BD Biosciences)
4. Reversible cell strainer (37 μm, STEMCELL Technologies)
5. Trypan blue stain 0.4% (Gibco BRL). Store at RT
6. Hemocytometer

2.9.2 Immunophenotypic Characterization of UCB CD34+-Enriched Cells

1. Centrifuge
2. FACS tubes (BD Biosciences)
3. PBS. Store at RT
4. Mouse anti-human monoclonal antibodies: IgG1, CD34, CD90, CD31, CD41, CD133, IgG2a, and CXCR4 (BioLegend)
5. PFA 1% (v/v). Store at 4 °C
6. FACSCalibur flow cytometer (BD Biosciences)
7. FACSFlow™ sheath fluid (BD Biosciences)

2.9.3 Clonogenic Assays (CFUs)

1. Cell culture incubator with CO_2, temperature, and humidity control
2. Inverted microscope
3. PBS. Store at RT
4. MethoCult GF H4434 (STEMCELL Technologies)
5. Nunc Thermo Fisher Multidish four wells

2.9.4 Cobblestone Area-Forming Cell (CAFC) Assay

1. Cell culture incubator with CO_2, temperature, and humidity control
2. Inverted microscope with phase contrast
3. Falcon® 24-well plate
4. MS-5 culture medium: DMEM 10% FBS 1% A/A
5. Mitomycin C (Sigma) (5 μg/mL prepared in DMEM 10% FBS)
6. DMEM 1% A/A
7. MyeloCult™ medium (STEMCELL Technologies)
8. Hydrocortisone 10^{-6} M (Sigma)

2.9.5 Adhesion Assay

1. Cell culture incubator with CO_2, temperature, and humidity control
2. Inverted microscope
3. Falcon® 96-well plate

4. PBS. Store at RT
5. Fibronectin (Sigma) (10 μg/mL)
6. DMEM 1% A/A
7. PFA 4% (v/v)
8. Crystal violet 0.5% (Sigma)
9. Distilled water

2.9.6 Migration Assay

1. Cell culture incubator with CO_2, temperature, and humidity control
2. Inverted microscope
3. Falcon® 24-well plate
4. PBS. Store at RT
5. 5 μm Pore polycarbonate transwell insert
6. Fibronectin (Sigma) (10 μg/mL)
7. BSA
8. DMEM supplemented with 1% A/A and 0.25% BSA
9. DMEM supplemented with 1% A/A, 0.25% BSA, and stromal cell-derived factor-1α (SDF-1α) (100 ng/mL)
10. Trypan blue stain 0.4% (Gibco BRL). Store at RT

3 Methods

3.1 Thawing and Expansion of Human BM-Derived MSC

1. Quickly thaw a cryogenic vial of human BM MSC (1 mL) stored in a liquid/vapor phase nitrogen tank in a water bath at 37 °C.
2. Dilute the cells 10× in 9 mL of thawing medium warmed to 37 °C.
3. Centrifuge the MSC at 300 × *g* for 5 min to remove the dimethylsulfoxide (DMSO)-containing supernatant.
4. Resuspend the cells in 5 mL of MSC culture medium.
5. Count the total number of viable cells using the Trypan blue exclusion method. Mix 10 μL of cell suspension with an equal volume of 0.4% Trypan blue (2× dilution) and insert 10 μL into a hemocytometer. Dead cells are stained in blue whereas live cells remain unstained.
6. Plate the thawed MSC at a density of 3–5 × 10^3 cells/cm^2 on 10 mL (T75) of culture medium. Incubate the cells at 37 °C in a 21% O_2 and 5% CO_2 atmosphere.
7. Every 3 days remove the exhausted medium and replace with fresh culture medium.
8. When cells reach a confluency of 70–80%, remove medium and wash the cells with phosphate-buffered saline (PBS).

9. Incubate the cells with 0.05% (w/v) trypsin (Life Technologies)—1 mM EDTA (Sigma) for 7 min at 37 °C (3 mL for a T75) to obtain a single cell suspension.
10. Observe the T-flasks under a microscope to ensure that the cells were completely detached.
11. Add twice the volume of culture medium to the harvested cells in the flasks and collect the cell suspension into a polypropylene tube. Wash the flasks once with culture medium.
12. Centrifuge the cells at 300 × *g* for 5 min.
13. Discard the supernatant and resuspend the cell pellet in culture medium. Use the Trypan blue exclusion method to count the number of viable harvested cells.
14. To expand the cultures further, repeat **steps 6–13**.

3.2 Thawing and Expansion of MS-5 Murine Stromal Cells

1. Quickly thaw a cryogenic vial of MS-5 murine stromal cells (DSMZ) (1 mL) stored in a liquid/vapor phase nitrogen tank in a water bath at 37 °C.
2. Expand MS-5 cells following **steps 2–13** of Subheading 3.1.

3.3 Preparation of Microwells

1. Prepare a solution of agarose 3% (w/v): weight 3 g of agarose and add it to a 250-mL glass bottle. Add 100 mL of deionized water and autoclave it at 121 °C for 20 min (*see* **Note 1**).
2. Close the lid and allow the solution to cool down to RT.
3. Prepare the agarose microwells on the day of the experiment by heating the agarose in the microwave to its melting point until it liquefies (100–150 °C) (*see* **Note 2**).
4. Add 500 μL of agarose to PDMS molds containing 400 μm × 400 μm-sized etched pyramidal profiles confined to 15.6 mm diameter circumferences and allow it to cool down for approximately 5 min (*see* **Note 3**).
5. After curing, peel off the agarose from the molds.
6. Gently place the microwell inserts in a 24-well plate using autoclaved tweezers. Make sure that the microwells are facing up. Press the insert down to avoid formation of air bubbles beneath the insert (*see* **Note 4**).
7. Wash the inserts twice with DMEM 1% A/A (*see* **Note 5**).
8. Add 500 μL of culture medium to the wells containing the inserts and centrifuge the plate at 3000 × *g* for 5 min (*see* **Note 6**).

3.4 Formation of BM MSC Spheroids

1. Prepare a suspension of MSC at a density of 1.2 × 10^6 cells/mL of culture medium (*see* **Note 7**).
2. Add 500 μL of the cell suspension to the microwell inserts and allow the cells to settle down for 5 min.

3. Centrifuge the 24-well plates at 200 × *g* for 10 min to force cell aggregation (*see* **Note 8**).
4. Incubate the cells at 37 °C for 18 h.

3.5 Preparation of a Feeder Layer of BM MSC

1. Seed 1.2 × 10^5 MSC/well on a 24-well plate 24 h before the co-culture with CD34$^+$-enriched cells is initiated (*see* **Note 9**).
2. Treat the feeder layers of MSC with mitomycin C (Sigma) (0.5 μg/mL prepared in DMEM 10% FBS) for 3 h at 37 °C to assure cell growth arrestment and prevent stromal cell detachment over culture time.
3. Wash the MSC feeder layer three times with DMEM 10% FBS.
4. Incubate the mitomycin C-treated feeder layer at 37 °C and 5% CO_2 in DMEM 10% FBS until use on the following day.

3.6 Isolation and Cryopreservation of UCB MNC

1. Collect HSPC from fresh UCB obtained after written informed parent's consent (*see* **Note 10**).
2. Transfer the UCB from the blood bag into a sterile bottle using a syringe connected to a 18G needle. Measure the blood volume.
3. Dilute the collected blood using a solution of PBS/EDTA (approximately 3× dilution).
4. Add 15 mL of Ficoll-Paque medium (1.077 g/mL; GE Healthcare) to a 50-mL Falcon tube.
5. Carefully layer 35 mL of the diluted blood sample onto the Ficoll-Paque medium solution (*see* **Note 11**).
6. Centrifuge at 300 × *g* for 30 min without brakes.
7. Using a Pasteur pipette, collect the buffy coat layer into a new 50-mL tube (*see* **Note 12**).
8. Dilute the buffy coat by adding 35 mL of PBS 2 mM EDTA solution to a 50-mL falcon tube containing 15 mL of buffy coat. Gently, mix the cells.
9. Centrifuge the cells at 500 × *g* for 10 min.
10. Discard the supernatant and resuspend the cells in 50 mL of PBS.
11. Count the cells using a Türk's solution, performing a 10× dilution with PBS prior to cell count.
12. Centrifuge the cells at 500 × *g* for 10 min.
13. Freeze the mononuclear cells (MNCs) in cryogenic vials using a solution of Recovery™ Cell Culture Freezing Medium (~50 × 10^6 cells/vial/mL of Recovery Medium). Store the samples at −80 °C overnight.
14. The following day, transfer the frozen vials to a liquid/vapor phase nitrogen container until further use.

3.7 $CD34^+$-Enrichment of UCB HSPC

1. Quickly thaw the frozen vials of MNCs in a water bath at 37 °C.
2. Resuspend the cells in thawing medium containing DNAse I in order to digest possible cell debris and prevent cell clumping (*see* **Note 13**).
3. Centrifuge the cells at 300 × *g* for 5 min and remove the supernatant.
4. Resuspend the pellet in 50 mL of MACS buffer (*see* **Note 14**).
5. Determine the cell number using the Trypan blue dye exclusion method (*see* **Note 15**).
6. Centrifuge the MNCs at 300 × *g* for 10 min.
7. Resuspend the cell pellet and add 300 μL of MACS buffer to up to 1×10^8 MNCs.
8. Add 100 μL of FcR blocking reagent for up to 1×10^8 MNCs, followed by addition of 100 μL of CD34 microbeads (CD34 MicroBead kit, Miltenyi) (*see* **Note 16**).
9. Mix well and incubate for 30 min at 4 °C protected from light.
10. Wash cells with 10 mL of MACS buffer and centrifuge at 300 × *g* for 10 min.
11. Remove the supernatant and resuspend the cells in 500 μL of MACS buffer (*see* **Note 17**).
12. Depending on the number of total cells, choose an appropriate MACS column and MACS separator (*see* **Note 18**).
13. Insert the column in the magnetic field of an adequate MACS separator and prime it by adding 3 mL of MACS buffer (*see* **Note 19**).
14. Place a 30-μm filter on the top of the column and apply the cell suspension onto the column.
15. Wash the column three times with 3 mL of MACS buffer (*see* **Note 20**).
16. Remove the column from the magnetic field and place it on a suitable collection tube (15 mL falcon tube for an LS column).
17. Add 5 mL of MACS buffer onto the column. Immediately flush out the magnetically labelled cells by firmly pushing the plunger into the column.
18. Determine the number of cells using the Trypan blue dye exclusion method.
19. Centrifuge the cells at 200 × *g* for 5 min and resuspend in culture medium.

3.8 Co-culture of Human UCB $CD34^+$-Enriched Cells and BM MSC

Establish co-cultures of $CD34^+$-enriched cells and MSC (2-D feeder layers or 3-D spheroids) in the culture medium at 37 °C and 5% CO_2 for 7 days. Seed each well of a 24-well plate with a starting $CD34^+$ cell density of 5×10^4 cells/mL of culture medium (total volume/well of 1.2 mL).

3.8.1 2-D MSC Monolayer and $CD34^+$ Cells

1. Remove the medium from the mitomycin C-treated feeder layer of MSC.
2. Add 6×10^4 $CD34^+$ cells/well of a 24-well plate in 1.2 mL of culture medium.
3. Incubate at 37 °C and 5% CO_2 for 7 days.

3.8.2 3-D MSC Spheroids and $CD34^+$ Cells in Suspension

1. Place a 37-μm reversible strainer (with the arrow pointed upward) on the top of a falcon tube.
2. Dislodge the spheroids from the inserts by gently pipetting with a 1000-μL pipette tip.
3. Pipette the dislodged spheroids through the strainer to remove single cells that did not aggregate into the spheroids. While single cells will flow through, the spheroids will remain on the filter.
4. Dispense 1 mL of medium on the well to dislodge any remaining spheroids. Pipette the spheroid suspension up and down three times before adding the spheroids to the top of the strainer. Repeat this washing step three times.
5. Observe the agarose inserts under a microscope to ensure that no remaining spheroids were left on the inserts. If necessary, repeat **step 4**.
6. Invert the strainer and place over a new conical tube.
7. Collect the spheroids by washing with 2–5 mL of culture medium per well harvested.
8. Allow the spheroids to settle by gravity on the bottom of the falcon tube.
9. Discard the supernatant and resuspend the spheroids in the culture medium.
10. Transfer 1.2×10^5 MSC to Ultralow Attachment 24-well plates in 700 μL of culture medium.
11. Add 6×10^4 $CD34^+$ cells in 500 μL of culture medium to the Ultralow Attachment well containing 1.2×10^5 MSC.
12. Incubate at 37 °C and 5% CO_2 for 7 days.

3.8.3 3-D MSC Spheroids and $CD34^+$ Cells in Microwells

1. Slowly remove 1 mL of medium from each well containing the preformed MSC spheroids in the agarose inserts.
2. Add 6×10^4 $CD34^+$ cells in 1.2 mL of culture medium on top of the microwells containing the MSC spheroids (*see* **Note 21**).
3. Incubate at 37 °C and 5% CO_2 for 7 days.

3.8.4 3-D (MSC + $CD34^+$) Spheroids in Microwells

1. Add 600 μL of culture medium to the agarose inserts.
2. Centrifuge the 24-well plate containing the inserts at 3000 × *g* for 5 min.

3. Prepare a cell suspension with 1×10^6 MSC and 1×10^5 CD34$^+$ cells per mL of medium.
4. Add 600 μL of the cell suspension to the microwell.
5. Allow the cells to settle for 5 min.
6. Centrifuge the 24-well plates containing the inserts at $200 \times g$ for 10 min.
7. Incubate the spheroids for 7 days at 37 °C and 5% CO_2.

3.9 Immunophenotypic and Functional Characterization of UCB CD34$^+$-Enriched Cells Co-cultured with BM MSC

After 7 days of culture, harvest the UCB expanded cells (HSPC). To separate the MSC spheroids and HSPC for further analysis:

1. Place a 37-μm reversible strainer (with the arrow pointed upward) on the top of a falcon tube.
2. Pipette the suspension of cells containing MSC spheroids and CD34$^+$ cells through the strainer. The CD34$^+$ cells will flow through while the MSC spheroids will be retained on the strainer.
3. Wash the culture wells three times with 1 mL of PBS. Pipette the washing solution through the strainer.
4. Invert the strainer and place over a new conical tube.
5. Collect the MSC spheroids by washing with 2–5 mL of PBS per harvested well.

3.9.1 Cell Counting and Viability

1. Use the Trypan blue exclusion method to determine the number of hematopoietic cells.
2. Calculate the fold increase (FI) in total nucleated cells (TNC) by dividing the harvested number of cells obtained at Day 7 by the number of cells seeded at Day 0.

3.9.2 Immunophenotypic Characterization of UCB CD34$^+$-Enriched Cells

Analyze the immunophenotype of CD34$^+$-enriched UCB HSPC before and after co-culture with BM MSC by staining the cells with a panel of antibodies against: CD34 and CD90 for stem/progenitor cells; the chemokine receptor CXCR4, CD31 (platelet endothelial cell adhesion molecule, expressed on the cell surface of endothelial cells and hematopoietic cells such as monocytes, platelets, neutrophils, natural killer cells, megakaryocytes, and some T cells), CD41 (megakaryocyte lineage), and CD133 (primitive hematopoietic stem and progenitor cells).

1. Centrifuge the cells at $300 \times g$ for 5 min and suspend the pellet in 800 μL of PBS.
2. Distribute the cells onto eight FACS tubes (100 μL each at a concentration of 1–10 × 10^6 cells/mL) labelled according to the surface marker to be analyzed: IgG1, CD34, CD90, CD31, CD41, CD133, IgG2a, and CXCR4.

3. Add 5 μL of antibody per FACS tube (or as recommended by manufacturer) and incubate the cells for 15 min at RT protected from light (*see* **Note 22**).
4. Resuspend the cells in 2 mL of PBS.
5. Centrifuge at 300 × *g* for 5 min and discard the supernatant.
6. Resuspend the cells in 500 μL of PFA 1%.
7. Analyze the samples by flow cytometry using a FACSCalibur. Collect a minimum of 10,000 events per sample and use an appropriate software (*p.e.*, FlowJo, FlowLogic) for data analysis.

3.9.3 Clonogenic Assays (CFUs)

1. To identify the formation of CFUs, as previously described by Andrade and colleagues [21], add 1 × 10^3 of fresh (Day 0) UCB $CD34^+$-enriched cells or 5 × 10^3 expanded cells (Day 7) in 2 mL of MethoCult GF H4434 (Stem Cell Technologies) (*see* **Note 23**).
2. Perform the colony-forming assays in triplicates by distributing the cells suspended in 2 mL of MethoCult among three wells of a four-well multidish plate (Nunc Thermo Fisher) (*see* **Note 24**).
3. Incubate the cells for 14 days at 37 °C and 5% CO_2.
4. After 14 days, count and evaluate the presence of colony-forming unit–granulocyte macrophage (CFU-GM), colony-forming unit–granulocyte, erythroid, macrophage, megakaryocyte (CFU-MIX), and burst-forming unit–erythroid (BFU–E) progenitors by microscopic examination of the colonies following the classification suggested in the Atlas of Human Hematopoietic Colonies [24].

3.9.4 Cobblestone Area-Forming Cell (CAFC) Assay

1. To identify the formation of CAFCs as previously described by Andrade and colleagues [21], seed MS-5 murine stromal cells in 24-well plates at a concentration of 2 × 10^4 cells/mL in DMEM 10% FBS.
2. Once the monolayer of MS-5 cells reaches confluency (3–4 days), treat it with mitomycin C (Sigma) for 3 h at a concentration of 5 μg/mL.
3. Carefully, wash the wells three times with DMEM 1% A/A to remove the mitomycin C-containing medium.
4. Add 1 × 10^3 of freshly isolated (Day 0 of experiment) or 2 × 10^3 of co-cultured $CD34^+$-enriched cells for 7 days (Day 7 of experiment) to the mitomycin C-treated confluent monolayer of MS-5 murine stromal cells in 1 mL of MyeloCult™ medium (STEMCELL Technologies) supplemented with 10^{-6} M hydrocortisone (Sigma). Prepare duplicates per experimental condition.
5. After 14 days, evaluate the number of cobblestone areas composed by more than five tightly packed cells under an inverted microscope with phase contrast.

3.9.5 Adhesion Assay (Adapted from Kadekar and Colleagues [23])

1. Coat 96-well plates with fibronectin (Sigma) (10 μg/mL diluted in PBS, volume of 200 μL/well) and incubate for 1 h at 37 °C.
2. At Day 0 and Day 7 of cell culture, plate 1 × 10^5 UCB cells/well in 200 μL of DMEM 1% A/A and allow the cells to adhere to the bottom of the plate for 1 h.
3. Gently rinse the cells three times with 200 μL of PBS.
4. Fix the cells with a solution of PFA 4% (A/A) for 20–30 min.
5. Stain the cells with 100 μL of crystal violet 0.5% (Sigma) at RT for 30 min.
6. Wash the wells three times with 200 μL of distilled water.
7. Image the wells using an inverted microscope.
8. Count the number of cells per optical field.

3.9.6 Migration Assay (Adapted from Kadekar and Colleagues [23])

1. Coat a 5-μm pore polycarbonate transwell insert with 200 μL of fibronectin (10 μg/mL in PBS) (Sigma) for 1 h at 37 °C.
2. Gently wash the transwell insert three times with DMEM 0.25% BSA (v/v).
3. At Day 0 and Day 7 of cell culture, add 1 × 10^5 UCB HSPC to the transwell insert suspended in 100 μL of DMEM 0.25% BSA (v/v).
4. Add 600 μL of DMEM 0.25% BSA (v/v) supplemented with 100 ng/mL of SDF-1α to the lower chamber of a 24-well plate. To account for spontaneous migration, include controls with medium without SDF-1α.
5. Carefully transfer the transwell insert to the SDF-1α-containing medium and allow the HSPC to migrate through the transwell for 4 h at 37 °C (*see* **Note 25**).
6. Collect and count the migrated cells using the Trypan blue exclusion method.

4 Notes

1. The agarose will not fully dissolve unless heat is applied. Loosen the lid before autoclaving the agarose solution.
2. At RT, the agarose solution solidifies. For repeated use, store the agarose at RT and reheat in a microwave. Loosen the lid while heating the agarose and immediately close it once the solution liquefies to maintain sterility.
3. Do not overfill the PDMS molds with agarose to avoid the formation of a convex meniscus over the top of the mold. This could allow the medium to rush under the agarose insert and disturb the formation of the spheroids.

4. Be careful when handling the tweezers to avoid damaging the agarose microwells.
5. Do not let the agarose microwells dry.
6. If the mold is not set at the bottom of the well, gently press the inserts down with the tweezers and repeat the centrifugation step. Avoid forces over 3200 × g to prevent agarose deformation.
7. Adjust cell numbers as necessary for spheroids of desired size. Overflowing the agarose microwells with cells can lead to poorly formed spheroids or result in formation of spheroids with nonuniform sizes.
8. Limit the movement of the plate as it could disrupt the integrity of the spheroids.
9. A cell seeding density of 1.2×10^5 MSC/well of a 24-well plate allows the cells to reach confluency on the following day.
10. Parent's informed consent should be obtained, in agreement with the law on the setting standards of quality and safety for the donation, procurement, testing, processing, preservation, storage, and distribution of human tissues and cells and with the approval of the Ethics Committee of the hospital.
11. Hold the 50 mL falcon tube at a 45° angle to limit mixing of the Ficoll-Paque and the diluted blood. Avoid disturbing the layers. Warm the Ficoll-Paque density gradient medium to 18–20 °C before use.
12. A small pellet of cells can be formed on the side of the tube. To collect the cells, gently "rub" them with the end of the pipette.
13. Use 9 mL of thawing medium per 50×10^6 of thawed MNCs.
14. To minimize the formation of cell clumps, keep the MACS buffer at 4 °C.
15. Due to the elevated number of cells usually obtained at this step, a 20× dilution in PBS is recommended for a more accurate estimation of the cell number.
16. It is important to follow the order of reagent addition.
17. The final cell concentration should not exceed 2×10^8 cells/mL.
18. LS columns have a capacity of approximately up to 2×10^8 cells.
19. The values herein indicated are for an LS column.
20. Do not let the column dry.
21. Dispense the cells by slowly pipetting down the wall of the well. This would avoid that the MSC spheroids are dislodged from the microwells.

22. Use the antibodies in appropriate concentrations and make sure that the concentrations of the antibody and respective isotype are matched.
23. When added to the solution of MethoCult, the cells should be suspended in a volume of approximately 100 μL but not lower than 100 μL—if needed, add PBS to complete the final volume. This would minimize the formation of air bubbles as MethoCult is a viscous solution.
24. Due to the viscosity of the MethoCult solution, only approximately 500 μL of solution (from the 2 mL total volume) can be added per well (three) of a 4-well plate. Add 1 mL of sterile water to the fourth well to help keeping a humidified environment in the 4-well plate.
25. Avoid generating bubbles when transferring the transwell insert to the SDF-1α-containing medium.

Acknowledgments

MHC acknowledges Fundação para a Ciência e a Tecnologia (FCT), Portugal, for granting PhD scholarship SFRH/BD/52000/2012. Funding received by iBB—Institute for Bioengineering and Biosciences from FCT (UID/BIO/04565/2013) and from Programa Operacional Regional de Lisboa 2020 (Project N. 007317) is acknowledged. The authors also acknowledge the funding received from Programa Operacional Regional de Lisboa 2020 through the project *PRECISE—Accelerating progress toward the new era of precision medicine* (Project N. 16394) and from FCT through the project *Design and operation of a prototype packed-bed reactor for the production of hematopoietic stem/progenitor cells* (PTDC/QEQ-EPR/6623/2014).

References

1. Schofield R (1978) The relationship between the spleen colony-forming cell and the haemopoietic stem cell. Blood Cells 4:7–25
2. Méndez-Ferrer S, Michurina TV, Ferraro F, Mazloom AR, MacArthur BD, Lira SA et al (2010) Mesenchymal and haematopoietic stem cells form a unique bone marrow niche. Nature 466(7308):829–834
3. Walenda T, Bork S, Horn P, Wein F, Saffrich R, Diehlmann A et al (2010) Co-culture with mesenchymal stromal cells increases proliferation and maintenance of haematopoietic progenitor cells. J Cell Mol Med 14 (1–2):337–350
4. Ema H, Takano H, Sudo K, Nakauchi H (2000) In vitro self-renewal division of hematopoietic stem cells. J Exp Med 192 (9):1281–1288
5. Zandstra PW, Conneally E, Petzer AL, Piret JM, Eaves CJ (1997) Cytokine manipulation of primitive human hematopoietic cell self-renewal. Proc Natl Acad Sci 94(9):4698–4703
6. Hosokawa K, Arai F, Yoshihara H, Iwasaki H, Nakamura Y, Gomei Y et al (2010) Knockdown of N-cadherin suppresses the long-term engraftment of hematopoietic stem cells. Blood 116(4):554–563

7. Wein F, Pietsch L, Saffrich R, Wuchter P, Walenda T, Bork S et al (2010) N-cadherin is expressed on human hematopoietic progenitor cells and mediates interaction with human mesenchymal stromal cells. Stem Cell Res 4 (2):129–139
8. Nakamura-Ishizu A, Okuno Y, Omatsu Y, Okabe K, Morimoto J, Uede T et al (2012) Extracellular matrix protein tenascin-C is required in the bone marrow microenvironment primed for hematopoietic regeneration. Blood 119(23):5429–5437
9. Nilsson SK, Johnston HM, Whitty GA, Williams B, Webb RJ, Dernhardt DT et al (2005) Osteopontin, a key component of the hematopoietic stem cell niche and regulator of primitive hematopoietic progenitor cells. Blood 106(4):1232–1239
10. Jing D, Fonseca A-V, Alakel N, Fierro FA, Muller K, Bornhauser M et al (2010) Hematopoietic stem cells in co-culture with mesenchymal stromal cells—modeling the niche compartments in vitro. Haematologica 95 (4):542–550
11. Cook MM, Futrega K, Osiecki M, Kabiri M, Kul B, Rice A et al (2012) Micromarrows—three-dimensional coculture of hematopoietic stem cells and mesenchymal stromal cells. Tissue Eng Part C Methods 18(5):319–328
12. Futrega K, Atkinson K, Lott WB, Doran MR (2017) Spheroid coculture of hematopoietic stem/progenitor cells and monolayer expanded mesenchymal stem/stromal cells in polydimethylsiloxane microwells modestly improves *in vitro* hematopoietic stem/progenitor cell expansion. Tissue Eng Part C Methods 23 (4):200–218
13. Isern J, Martín-Antonio B, Ghazanfari R, Martín AM, López JA, del Toro R et al (2013) Self-renewing human bone marrow mesenspheres promote hematopoietic stem cell expansion. Cell Rep 3(5):1714–1724
14. Jeon S, Lee H-S, Lee G-Y, Park G, Kim T-M, Shin J et al (2017) Shift of EMT gradient in 3D spheroid MSCs for activation of mesenchymal niche function. Sci Rep 7(1)
15. Schmal O, Seifert J, Schaffer TE, Walter CB, Aicher WK, Klein G (2016) Hematopoietic stem and progenitor cell expansion in contact with mesenchymal stromal cells in a hanging drop model uncovers disadvantages of 3D culture. Stem Cells Int 2016:4148093
16. Greenbaum A, Hsu Y-MS, Day RB, Schuettpelz LG, Christopher MJ, Borgerding JN et al (2013) CXCL12 in early mesenchymal progenitors is required for haematopoietic stem-cell maintenance. Nature 495(7440):227–230
17. Singh H, Mok P, Balakrishnan T, Rahmat SNB, Zweigerdt R (2010) Up-scaling single cell-inoculated suspension culture of human embryonic stem cells. Stem Cell Res 4 (3):165–179
18. Banerjee M, Bhonde RR (2006) Application of hanging drop technique for stem cell differentiation and cytotoxicity studies. Cytotechnology 51(1):1–5
19. Dontu G, Abdallah WM, Foley JM, Jackson KW, Clarke MF, Kawamura MJ et al (2003) In vitro propagation and transcriptional profiling of human mammary stem/progenitor cells. Genes Dev 17(10):1253–1270
20. Lewis NS, Lewis EEL, Mullin M, Wheadon H, Dalby MJ, Berry CC (2017) Magnetically levitated mesenchymal stem cell spheroids cultured with a collagen gel maintain phenotype and quiescence. J Tissue Eng 8:1–11
21. Andrade PZ, dos Santos F, Almeida-Porada G, Lobato da Silva C, S. Cabral JM. (2010) Systematic delineation of optimal cytokine concentrations to expand hematopoietic stem/progenitor cells in co-culture with mesenchymal stem cells. Mol BioSyst 6(7):1207
22. da Silva CL, Gonçalves R, Crapnell KB, Cabral J, Zanjani ED, Almeida-Porada G (2005) A human stromal-based serum-free culture system supports the ex vivo expansion/maintenance of bone marrow and cord blood hematopoietic stem/progenitor cells. Exp Hematol 33(7):828–835
23. Kadekar D, Kale V, Limaye L (2015) Differential ability of MSCs isolated from placenta and cord as feeders for supporting ex vivo expansion of umbilical cord blood derived CD34+ cells. Stem Cell Res Ther 6:201
24. Eaves C, Lambie K (1995) Atlas of human hematopoietic colonies. STEM CELL Technol Inc., Vancouver, BC

Methods in Molecular Biology (2019) 2002: 121–128
DOI 10.1007/7651_2018_182

Published online: 22 September 2018

Investigating the Vascular Niche: Three-Dimensional Co-culture of Human Skeletal Muscle Stem Cells and Endothelial Cells

Claire Latroche, Michèle Weiss-Gayet, and Bénédicte Chazaud

Abstract

Angiogenesis, the growth of new blood vessels, is crucial for efficient skeletal muscle regeneration. Myogenesis and angiogenesis take place concomitantly during muscle regeneration. Myogenic precursor cells (MPCs) are in close proximity to vessels and interact with neighboring endothelial cells (ECs) to expand and differentiate. To demonstrate functional interplay between the two cell types, we established a robust and predictive ex vivo assay to evaluate activity of MPCs on angiogenesis and vice-et-versa, of ECs on myogenesis. Here, we describe an optimized three-dimensional co-culture protocol for the assessment of biological interactions between MPCs and ECs during skeletal muscle regeneration.

Keywords 3D tri/co-culture, Angiogenesis, Endothelial cells, Fibrin gel, Muscle stem cells, Myogenesis, Niche

1 Introduction

Vascular stem cell niche has been well characterized in several tissues such as bone marrow for hematopoietic stem cells and central nervous system for neural stem cells [1, 2]. Within these tissues, vessels are always associated with stem cells and actively participate to both quiescence maintenance and proliferation and differentiation at the time of tissue repair. While the vascular niche concept has been also raised in skeletal muscle, it remains poorly characterized. Satellite cells (SCs) are the principal players in skeletal muscle regeneration. Upon injury, quiescent SCs activate and become MPCs that expand and differentiate to finally fuse together, giving rise to new functional myofibers. A subset of MPCs does not differentiate and self-renews to replenish the SC pool [3]. At the quiescent state, SCs are located under the basal lamina, along the myofiber, close to vessels [4, 5]. Muscle regeneration includes myogenesis and concomitant angiogenesis [6, 7]. Relationship between ECs and SCs/MPCs has been established in vitro and in vivo [6, 8–10] and remains to be fully characterized at the molecular level.

Angiogenesis is a complex multistep process including the degradation of the basement membrane, migration, proliferation, and alignment of ECs into the extracellular matrix to undergo tube formation with lumen. Eventually, neo-tubes branch and anastomose with adjacent vessels. We optimized an in vitro angiogenesis assay from an original assay [11] that recapitulates all the key stages of angiogenesis and, importantly, allows the formation of fully differentiated vessels exhibiting lumens. Human umbilical vascular cells (HUVEC) or human dermal microvascular endothelial cells (HDMEC) are coated onto Cytodex microcarriers and embedded into a fibrin gel. Using this technique, specific interactions between MPCs and ECs were studied by seeding MPCs either on the top of the gel where they provide soluble factors to ECs or into the gel where they can establish direct cell–cell interactions with ECs. The later technique requires previous fluorescent labeling of cells to distinguish between ECs and MPCs in the analysis. A third cell partner can be added to the setup such as macrophages or fibroblasts to analyze their impact of myo/angiogenesis. This tri/coculture may be used for the direct functional analysis of specific effectors (blocking antibodies, recombinant proteins, silencing, etc., as in [10]), or to study specific diseases (specific mutation of one cell type, cells extracted from diseased tissues, etc.).

2 Materials

2.1 Cytodex 3 Beads

Cytodex beads (Amersham Pharmacia, 17-0485-01) reconstitution: 0.5 g of dry beads are hydrated and swollen in 50 ml PBS (pH = 7.4) under agitation for at least 3 h at room temperature (RT). Use a 50-ml tube and place it on the rocker. Let the beads settle down (about 15 min) (*see* **Note 1**). Discard the supernatant and wash the beads for a few minutes in PBS 1× (50 ml). Discard PBS and replace with 50 ml PBS 1× in order to have 30,000 beads/ml. Pour the bead suspension in a siliconized glass bottle (Windshield Wiper or Sigmacote). Sterilize the beads by autoclaving for 15 min at 115 °C. Store at 4 °C.

2.2 Fibrinogen Solution

Fibrinogen (Sigma Aldrich, F-8630) dissolution: 2.5 mg/ml fibrinogen in Endothelial Cell Basal Medium 2 (ECBM2, Promocell, C22-210). Heat in a 37 °C water bath to dissolve the fibrinogen. Mix by inverting the tube. Do not vortex. Sterilize through 0.22 μm filter.

2.3 Aprotinin

Aprotinin (Sigma Aldrich, A-1153) reconstitution: homogenize lyophilized aprotinin at 4 U/ml in sterile ultrapure water. Sterilize through 0.22 μm filter and store at −20 °C in 0.5 ml aliquots.

2.4 Thrombin

Thrombin (Sigma Aldrich, T-9549) reconstitution: homogenize in sterile ultrapure water at 100 U/ml. Aliquots of 50 μl are stored at −80 °C.

2.5 Cell Media

1. HUVEC are cultured in ECGM2: Endothelial Cell Growth Medium 2 (Promocell C22011 that is supplemented with mix containing growth factors).
2. HDMEC cells are cultured in ECGMV2: Endothelial Growth Medium V2 (Promocell C22-022 that is supplemented with mix containing growth factors).
3. Human MPCs and fibroblasts are cultured in HAMF12 medium (Gibco 31765) containing 15% FBS (Gibco 10270) and 10,000 U penicillin/streptomycin (Gibco 15140).
4. Anti-inflammatory macrophages are cultured in Advanced RPMI 1640 medium (Gibco 12633) supplemented with 1× Glutamax (Gibco 38050), 10,000 U penicillin/streptomycin (Gibco 15140), 10 mM HEPES (Gibco 15630), 1× MEM vitamins (Gibco 11120), and 50 μM 2-mercaptoethanol (Gibco 31350), 0.5% FBS (Gibco 10270) supplemented with either interleukin-4 (IL-4) (10 ng/ml) (R&D 204-IL/CF) or IL-10 (10 ng/ml) (R&D 217-IL/CF) + dexamethasone (80 ng/ml) (Sigma D4902).
5. After seeding on the top of the gel, cells were cultured in EndoGro medium (Millipore, SCME-BM that is supplemented with mix containing growth factors).

2.6 Culture Supports

1. T75 flasks are used for human MPCs, fibroblasts, or HUVEC/HDMEC growing and coated on beads.
2. Eight well chamber glass slides (Lab-Tek Nunc 177402) are used for 3D co-cultures.
3. PFTE inserts for six-well plates (Millipore PICM0350) are used for human macrophage culture.

3 Methods

3.1 Preparing Cells

1. Culture HUVEC or HDMEC (in ECGM2 or ECGMV2 medium, respectively) a few days before seeding onto the beads (*see* **Note 2**). A concentration of about 400 cells per bead is required.
2. Macrophages (260,000 cells per insert) are activated with IL-4 or with IL-10 + dexamethasone for 2 days as described in [12] before co-culture. 60,000 Macrophages per well are needed.

3. Human fibroblasts or MPCs (*see* **Note 3**) are cultured in HAMF12 medium and switched to EndoGro medium one day before seeding on fibrin. 30,000 Cells per well are needed.

3.2 Coating the Beads with HUVEC

1. Trypsinate non-confluent HUVEC/HDMEC.
2. In parallel, allow beads to settle down (*see* **Note 1**), aspirate the supernatant, and wash the beads briefly in 2 ml of warm ECGM2 medium.
3. To fill 2 Lab-Teks (i.e., 16 wells), mix 1500 beads (*see* **Note 4**) with 60,000 HUVEC in 2 ml of warm ECGM2 medium in a 5-ml round tube (polypropylene) and place it vertically in the incubator.
4. Incubate for 4 h at 37 °C, shaking softly the tube every 15 min (*see* **Note 5**) during 2 h, then every 30 min.
5. Transfer the coated beads into a T75 flask in 12 ml of ECGM2 medium and incubate overnight (*see* **Note 6**).

3.3 Embedding Coated Beads in Fibrin Gel

1. Prepare the eight-well chamber Lab-Tek by incubation at 37 °C (*see* **Note 7**).
2. Prepare the 2.5 mg/ml fibrinogen solution in ECBM2 medium (*see* **Note 8**) and shake thoroughly.
3. Add aprotinin 0.15 U/ml (500×) to the fibrinogen solution and sterilize through a 0.22-μm filter.
4. Transfer coated beads into a 15-ml conical tube and let beads settle down (*see* **Note 9**).
5. Wash the beads three times with 5–10 ml of ECBM2 medium (*see* **Note 10**).
6. Optional (*see* **Note 11**). Count beads in a defined volume (e.g., 10 μl) on a coverslip.
7. Homogenize the beads in the fibrinogen solution at a concentration of about 250 beads/ml.
8. Optional. For cell–cell interaction study, mix the fibrinogen solution and beads with RFP-MPCs (*see* **Note 3**).
9. Prepare a 10-U/ml thrombin solution in ECBM2 medium (stock is 100 U/ml). Add 15 μl to each well (*see* **Note 12**).
10. Add 0.3 ml of the fibrinogen/bead suspension to one well.
11. Mix the thrombin and the fibrinogen by going up and down gently with the pipette tip approximately 4–5 times (*see* **Note 13**).
12. Repeat **steps 10–11** for each well (*see* **Note 14**).
13. Leave the Lab-Tek for 2 min under the hood and then place it in the 37 °C incubator for 10–15 min to generate a clot (*see* **Note 15**).

14. Seed cells to be tested (30,000 fibroblasts, MPCs, or 60,000 macrophages) in 0.2 ml of EndoGro medium on top of the fibrin gel (*see* **Note 16** and Fig. 1).
15. Change the media every 2 days (EndoGro medium) (*see* **Note 17**).

3.4 Recording and Analysis

1. At day 6, wash three times with PBS 1× for 5 min and fix the cells with 4% PFA for 30 min at RT, then wash three times with PBS 1× for 5 min.
2. Nuclei are stained by adding Hoechst (1/1000) for 15 min.
3. Wash with PBS 1× and store at 4 °C (*see* **Note 18**) until recording.
4. Samples are observed using a Zeiss Axio Observer.Z1 microscope and recorded at 10× magnification with a Photometrics CoolSNAP HQ2 camera using MetaView software (*see* **Note 19**).
5. Quantification is performed using ImageJ. Sprouts are measured as tubes originating from the Cytodex bead. Each sprout length is calculated. The presence of lumen in the sprout is quantified as well as the number of myotubes in the gel. For each condition, a minimum of 100 sprouts is analyzed (Fig. 1).

4 Notes

1. It is not necessary to centrifuge, as the beads are heavy enough to settle down. Let them about 15 min after each wash before discarding the supernatant (whether they are coated with ECs or not).
2. The experiment can be run with unlabeled HUVEC. Sprouts are visible under the light microscope. However, it is easier, notably in the co-culture, to label the cells with a fluorescent dye. HUVEC are easily infected with lenti-GFP at early passage and then purified by cell sorting before expansion.
3. When MPCs are embedded into the gel, it is preferable to label them before the co-culture. Human MPCs are easily infected by lentiviruses. MPCs at passage 1 are incubated with lenti-RFP in growing medium without antibiotics overnight. The medium is changed and cells are grown for 3 days before use. Each batch of lentivirus must by tittered to use the appropriate concentration to reach about 100% of infection.
4. To manipulate beads, use large end p200 tips (inside diameter of at least 1.5 mm) or cut smaller tips in a sterile way.
5. To avoid EC take off from the beads, gently shake or vortex at very low speed the tube.

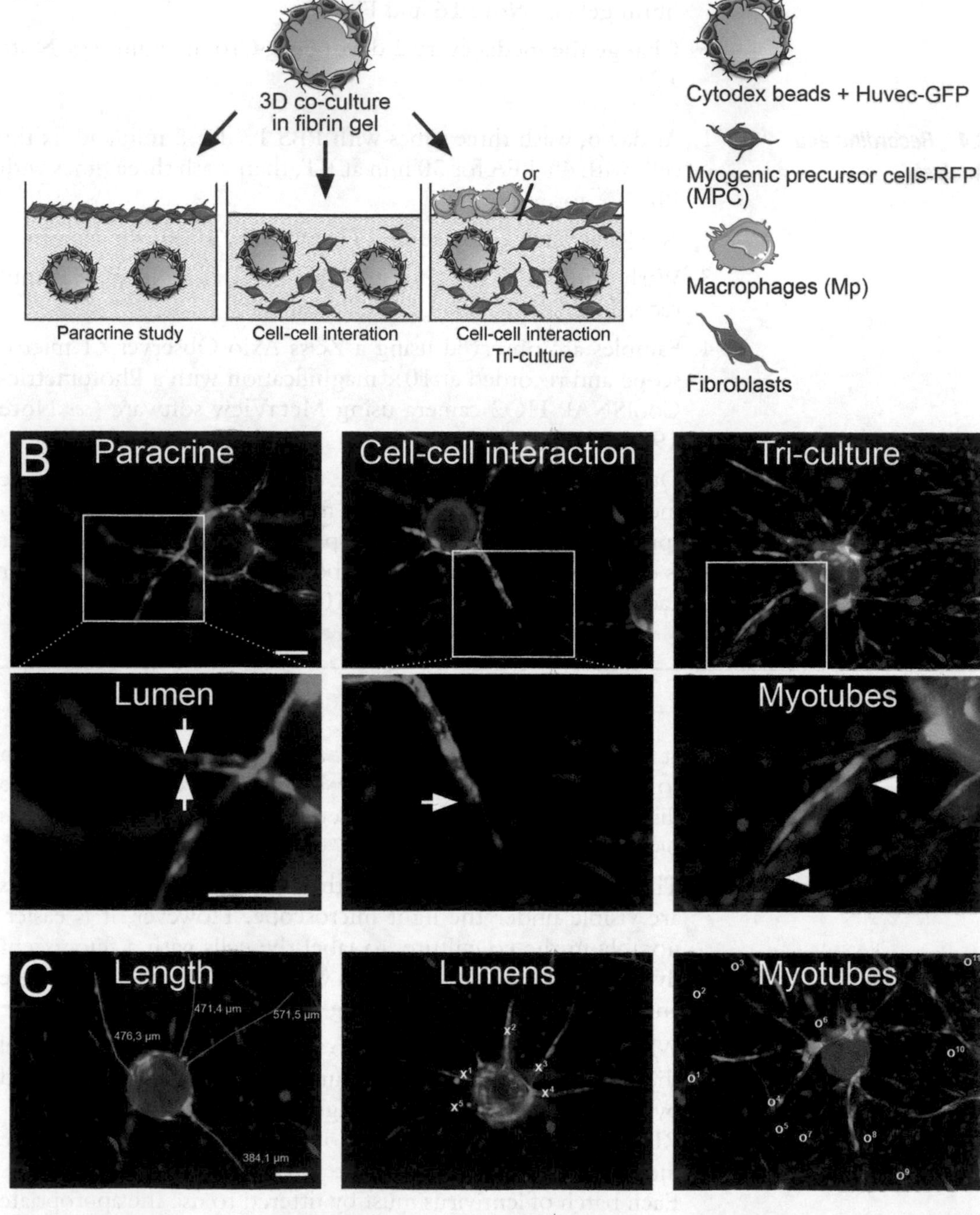

Fig. 1 Three-dimensional co-cultures of ECs and different cell partners. (**a**) Scheme explaining various co-culture conditions: beads recovered with ECs can be seeded in 3D with cells on the top of the gel to study their paracrine effect (left), inside the gel to study direct cell:cell interactions (middle), or in tri-culture with various cell types both at the top and inside the gel (right). (**b**) Images representing the various conditions: paracrine (left), direct cell–cell interactions (middle), and tri-culture (right). Each image is zoomed to appreciate lumen formation (arrows) and myotubes (arrowheads). (**c**) Quantification examples for length (white line), lumens number (x), and the number of myotubes (o). Examples are given with GFP-HUVEC and RFP-MPCs. Bars: 100 μm

6. Check coated beads under the microscope, which should look like mini golf balls after beading.
7. One may work with 48-well plates. However, Lab-Tek glass slides are manufactured for excellent optical quality.
8. From this point, it is important to work in basal medium, without any FBS. The presence of FBS would block the reaction between fibrinogen and thrombin.
9. Rinse the T75 flask with ECGM2 medium, and slowly pipet beads to avoid detachment of ECs from beads.
10. It is important to wash FBS to allow fibrin gel formation.
11. Counting is useful when beads have to be split in various conditions.
12. Seed the 15 μl of thrombin solution in the middle of the well; it will be easier than to mix with fibrinogen.
13. Be careful not to make bubbles.
14. Change the pipette tip for each well.
15. Usually, when the fibrin gel is formed, some tiny bubbles are visible in the gel, which will disappear in 3–4 days.
16. Pour cells drop by drop. This step can be done either on Day 0 or Day 1.
17. By day 3 or 4, beginning of sprouting should be visible.
18. Add PBS 1×, 0.05% sodium azide to avoid drying of wells and contamination.
19. Since sprouts are not in horizontal alignment, several pictures must be recorded for the same bead, in order to measure the whole length of the sprout. Similarly, myotubes are located in the whole gel and several pictures at different focus along the *z* axis must be recorded.

Acknowledgments

We thank Stéphane Germain and Laurent Muller (CIRB, Collège de France, Paris) for the transfer of the 3D angiogenesis technique in our lab. B. Chazaud group research using this technique has been recently supported by INSERM, CNRS, Université Claude Bernard Lyon I, and grants from the Framework Programme FP7 Endostem (grant agreement 241440), and Association Française contre les Myopathies (grants 18003 and 111199). C.L. was supported by Dim Stem Pole from Region Ile-de-France and Association Française contre les Myopathies.

References

1. Mendelson A, Frenette PS (2014) Hematopoietic stem cell niche maintenance during homeostasis and regeneration. Nat Med 20:833–846
2. Nakagomi N, Nakagomi NT, Kubo S, Nakano-Doi A, Saino O, Takata M et al (2009) Endothelial cells support survival, proliferation, and neuronal differentiation of transplanted adult ischemia-induced neural stem/progenitor cells after cerebral infarction. Stem Cells 27:2185–2195
3. Yin H, Price F, Rudnicki MA (2013) Satellite cells and the muscle stem cell niche. Physiol Rev 93:23–67
4. Mounier R, Chrétien F, Chazaud B (2011) Blood vessels and the satellite cell niche. Curr Top Dev Biol 96:121–138
5. Latroche C, Gitiaux C, Chrétien F, Desguerre I, Mounier R, Chazaud B (2015) Skeletal muscle microvasculature: a highly dynamic lifeline. Physiology 30:417–427
6. Christov C, Chretien F, Abou-Khalil R, Bassez G, Vallet G, Authier FJ et al (2007) Muscle satellite cells and endothelial cells: close neighbors and privileged partners. Mol Biol Cell 18:1397–1409
7. Abou-Khalil R, Mounier R (2010) Chazaud B (2010) Regulation of myogenic stem cell behaviour by vessel cells: the "ménage à trois" of satellite cells, periendothelial cells and endothelial cells. Cell Cycle 9:892–896
8. Renault MA, Vandierdonck S, Chapouly C, Yu Y, Qin G, Metras A et al (2013) Gli3 regulation of myogenesis is necessary for ischemia-induced angiogenesis. Circ Res 113:1148–1158
9. Rhoads RP, Johnson RM, Rathbone CR, Liu X, Temm-Grove C, Sheehan SM et al (2009) Satellite cell-mediated angiogenesis in vitro coincides with a functional hypoxia-inducible factor pathway. Am J Physiol Cell Physiol 296:C1321–C1328
10. Latroche C, Weiss-Gayet M, Muller L, Gitiaux C, Leblanc P, Liot S et al (2017) Coupling between myogenesis and angiogenesis during skeletal muscle regeneration is stimulated by restorative macrophages. Stem Cell Rep 9:2018–2033
11. Nakatsu MN, Davis J, Hughes CCW (2007) Optimized fibrin gel bead assay for the study of angiogenesis. J Vis Exp. https://doi.org/10.3791/186
12. Saclier M, Yacoub-Youssef H, Mackey AL, Arnold L, Ardjoune H, Magnan M et al (2013) Differentially activated macrophages orchestrate myogenic precursor cell fate during human skeletal muscle regeneration. Stem Cells 31:384–396

Methods in Molecular Biology (2019) 2002: 129–139
DOI 10.1007/7651_2018_184

Published online: 30 September 2018

Identification and Characterization of Stem Cells in Oral Cancer

Sujit K. Bhutia, Prajna P. Naik, Prakash P. Praharaj, Debasna P. Panigrahi, Chandra S. Bhol, Kewal K. Mahapatra, Sarbari Saha, and Srimanta Patra

Abstract

Cancer stem cells (CSCs) are a subpopulation of cells within a heterogeneous tumor that have enhanced biologic properties such as increased capacity for self-renewal, increased tumorigenicity, enhanced differentiation capacity, and resistance to chemo- and radiotherapies. This unit describes protocols to isolate and characterize potential cancer stem cells from a solid tumor (oral cancer). This involves creating a single-cell suspension from tumor tissue, tagging the cell subpopulation of interest, and sorting cells into different populations. Finally, the sorted subpopulations can be evaluated for their ability to meet the functional requirements of a CSC, which primarily include increased tumorigenicity in an in vivo xenograft assay. Mastering the protocols in this unit will allow the researcher to study populations of cells that may have properties of CSCs.

Keywords Aldehyde dehydrogenase activity, Cancer stem cells, CD44, Orosphere, Side population, Xenograft

1 Introduction

The failure of the current therapies to cure the oral squamous cell carcinoma is due their recurrence property, chemo–radio-resistance, metastasis which are manifested by subgroup of cancer cells termed as cancer stem cells (CSCs) [1]. The long-term self-renewal property of these CSCs is due to the activation of the various developmental signaling pathways such as Notch, Wnt, and Hedgehog, which are also primarily involved in the self-renewal of normal stem cells [2]. The fundamental changes in the processes that are found to be involved in the resistance property of oral CSCs include: aberration in Hedgehog and Notch signaling pathways, altered drug responses, alteration in cell-cycle regulation, anti-apoptosis, autophagy, altered metabolic pathways, faulty DNA damage response, and defective epigenetic modifications [1]. It should be noted that there is no specific marker that is universally accepted as a CSC marker. Even for individual tumor types, a certain population may be described as having stem cell properties but cannot conclude that it is the only, or the most reliable, CSC populations. The differential characteristics of the CSCs can be employed to get themselves to be sorted using several techniques, which include: sphere formation assay, surface markers-based sorting by flow cytometry (CD133 CD44), ALDEFLUOR assay, and Hoechst dye exclusion assay. Furthermore, the functional characterization can be done by xenotransplantation assay, which assesses the in vivo self-renewal and tumorigenic potential of the CSCs [1]. The sphere formation is the most predominant assay that helps in the sorting of the CSCs in low serum media and nonadherent plates. The formation of solid spheres after some days of incubation facilitates the separation of CSCs as only CSCs are specialized to grow under such conditions [3]. The presence of the distinct surface markers on CSCs (CD133, CD44, CD90, CD105, etc.) can be used to segregate them using flow cytometry which provides an effective method to separate different types of CSCs. Additionally, the higher expression of ABC transporters in CSCs provides a flow cytometry analysis of such cells as side populations. These side population cells show no fluorescence with DNA-binding dyes like Hoescht33342 in flow cytometry analysis. The low uptake of the dyes is supplemented with the presence of higher ABC transporters which helps in the characterization of CSCs [4]. Similarly, the alteration in the cell-cycle/G2-M arrest and anti-apoptosis nature of the CSCs can be determined with flow cytometer along with the ALDEFLUOR assay, which shows high fluorescence with Aldefluor reagent due to the presence of high ALDH1 activity. Thus, the higher fluorescence is proportional to the high ALDH1 expression which validates the density of CSCs with the help of flow cytometer [5]. This chapter therefore tries to document the different characterization procedures of the CSCs, which can be an effective method to sort out their subgroups and provide an easy therapeutic target.

2 Materials

2.1 Dissociation of Primary Tumor or Mouse Xenografts

1. Tumor sample (preferred at least 1 cm of viable tissue)
2. RMPI-1640 (HIMEDIA, cat. no. AL028A) or DMEM/F12 (1:1 mixture) (HIMEDIA, cat. no. AL140) cell culture media (or preferred media)
3. 100 mm Petri dish (Sigma, cat. no. EP0030702115); dissection tools (NeoLab)
4. 50 mL Conical tube 70-μm sterile mesh filter (Sigma, cat. no. CLS431751)
5. 10, 5, and 1 mL Serological pipettes, 16-gauge needles, and 3–5 mL syringes

6. Hemocytometer (Hausser Scientific Bright-Line™ Counting Chamber, cat. no. 02-671-51B); Trypan blue (Sigma, cat. no. T8154)
7. Enzymatic digestion solution [TrypLE Express (Life Technologies, cat. no. 12605-010)]
8. Collagenase type II (Life Technologies, cat. no. 17101-015): Dissolve 5 mg of collagenase type II in 1 mL of RPMI-1640/DMEM-F12 to make a 5-mg/mL solution

2.2 Analysis of CSCs in Cancer Cell Lines or Primary Tumor or Mouse Xenografts

2.2.1 Surface Markers: CD44+

1. Phosphate buffer saline (PBS): Prepare 1× phosphate buffer saline consisting of 0.39 g NaH_2PO_4 (sodium phosphate monobasic) (Sigma, cat. no. S5011), 1.45 g Na_2HPO_4 (sodium phosphate dibasic) (Sigma, cat. no. V800397), and 9.02 g NaCl (sodium chloride) (Sigma, cat. no. S6191). Maintain pH 7.4. Store it at 4 °C
2. DMEM (HIMEDIA, cat. no. AL151A)
3. FBS (Gibco™, cat. no. 10270106)
4. Trypsin–EDTA (HIMEDIA, cat. no. TCL007)
5. Stem cells surface marker: CD44-FITC (BD Biosciences, cat. no. 347943), and CD133-PE (BD Biosciences, cat. no. 566593)
6. Equipment: flow cytometer

2.2.2 The Side Population: Hoechst Dye Exclusion

1. PBS 1× (0.39 g NaH_2PO_4 (Sigma, cat. no. S5011), 1.45 g Na_2HPO_4 (Sigma, cat. no. V800397), and 9.02 g NaCl (Sigma, cat. no. S6191))
2. DMEM (HIMEDIA, cat. no. AL151A)
3. FBS (Gibco™, cat. no. 10270106)
4. Hoechst 33342 (Sigma, cat. no. B2261)
5. Trypsin–EDTA (HIMEDIA, cat. no. TCL007)
6. Equipment: flow cytometer

2.2.3 Enzymatic Activity: ALDEFLUOR Assay

1. PBS 1× (0.39 g NaH_2PO_4 (Sigma, cat. no. S5011), 1.45 g Na_2HPO_4 (Sigma, cat. no. V800397), and 9.02 g NaCl (Sigma, cat. no. S6191))
2. DMEM (HIMEDIA, cat. no. AL151A)
3. FBS (Gibco™, cat. no. 10270106)
4. ALDEFLUOR kit (Stemcell Technologies, Vancouver, CA, cat. no. 01700)
5. Trypsin–EDTA (HIMEDIA, cat. no. TCL007)
6. Equipment: flow cytometer

2.2.4 Cell Culture Selection: Orosphere Assay

1. B-27® Supplement (50×) serum free (Gibco™, cat. no. 17504-044)
2. Recombinant Fibroblast Growth Factor-basic (bFGF-2) (Gibco™, cat. no. PHG0264)
3. Recombinant Epidermal Growth Factor (EGF) (BD Biosciences, cat. no. 354001)
4. DMEM (HIMEDIA, cat. no.AL151A)
5. PBS 1× (0.39 g NaH_2PO_4 (Sigma, cat. no. S5011), 1.45 g Na_2HPO_4 (Sigma, cat. no. V800397), and 9.02 g NaCl (Sigma, cat. no. S6191))
6. Ultralow attachment multiple well plates (Costar®, cat. no. 3471)
7. Accutase solution (Sigma, cat. no. A6964); trypsin–EDTA (HIMEDIA, cat. no. TCL007)
8. Equipment: (Olympus IX71 fluorescent inverted microscope and the cellSens standard software (version 1.6, Olympus Soft Imaging Solutions GmbH, Johann-Krane-Weg, Münster, Germany)

2.3 Xenograft Formation Assay

1. Six- to eight-week-old SCID mice of desired gender (kept in approved IACUC and ARP conditions)
2. Putative sorted cancer stem cell population and non-cancer stem cell population
3. Isoflurane
4. Vaporizing anesthesia device
5. 1 mL Syringe; 25-gauge needles
6. Matrigel® Growth Factor Reduced (GFR), phenol red-free (BD Biosciences, cat. no. 356231)
7. DMEM (HIMEDIA, cat. no.AL151A) and betadine scrub

3 Methods

3.1 Dissociation of Primary Tumor or Mouse Xenografts

In this section, the detailed protocol for mechanical/chemical dissociation (*see* **Note 1**) of a tumor tissue sample that comes from either a human patient or a mouse xenograft is described.

1. Process the tumor sample as soon as possible to maximize viability. Ideally, process the tissue within 30 min of removal. *See* **Note 2**.
2. After receiving the tissue sample from a patient or removing a tumor from a mouse, place the specimen in a 100-mm cell culture disc with approximately 2 mL of ice-cold media with

collagenase (5 mg/mL). Place the disc on ice during dissection.

3. Hold the specimen firm with tissue forceps, and scrape the specimen downward, and away using a number 22 scalpel blade, such that cells are pulled off the tumor mass into the disc.
4. Continue scrapping until the specimen is too small to hold and you have a large "slurry" population in the 100-mm disc.
5. Transfer the slurry to a 50-mL conical tube and add 10 mL of media (DMEM/F12) with collagenase (5 mg/mL) and incubate at 37 °C for 1 h.
6. Further dissociate the tumor sample by adding 10 mL of 0.25% TrypLE E with or without hyaluronidase and incubate at 37 °C for 15 min. At every 5-min interval, pipette up and down the solution using a 10-mL serological pipette, followed by 5 and 1 mL serological pipette until it passes freely.
7. Neutralize the TrypLE E-cell solution with 20 mL of DMEM medium with 10% FBS (or 10 mL in a 15-mL conical tube).
8. Use a 5-mL pipette to pass the cell suspension through a 70-μm sterile mesh filter to generate a single-cell suspension. Slowly deposit the cell suspension onto the filter and allow it to pass into the collection tube. *See* **Note 3**.
9. After collecting the entire cell suspension, use 50 μL of it to count and assess the cell viability using Trypan blue exclusion. Centrifuge the remaining sample at 3000 rpm (1500 × *g*) for 10 min at 4 °C.
10. Discard the supernatant and resuspend the cell pellet in appropriate volume of culture media or PBS based on cell density calculated by Trypan blue exclusion.

3.2 Analysis of CSCs in Cell Lines or Primary Tumor or Mouse Xenografts

3.2.1 Surface Markers: CD44+

1. Prepare single viable cell suspension from established cancer cell lines or from Subheading 3.1 as per the experimental setup (about 1×10^6 cells for analysis and 1×10^7 cells for sorting).
2. Wash the cells with 1× sterile PBS as required. *See* **Note 4**.
3. Pellet down and resuspend the cells with 200 μL of PBS.
4. Stain the cells with 5 μL CD44 antibody and CD133 per 1×10^6 cells along with isotypes. *See* **Note 5**.
5. Vortex gently and incubate for 30 min at room temperature in dark.
6. Wash the cells with PBS (*see* **Note 4**) and resuspend it with 500 μL PBS.
7. Acquire the cell population with flow cytometry (Fig. 1). *See* **Note 6**.

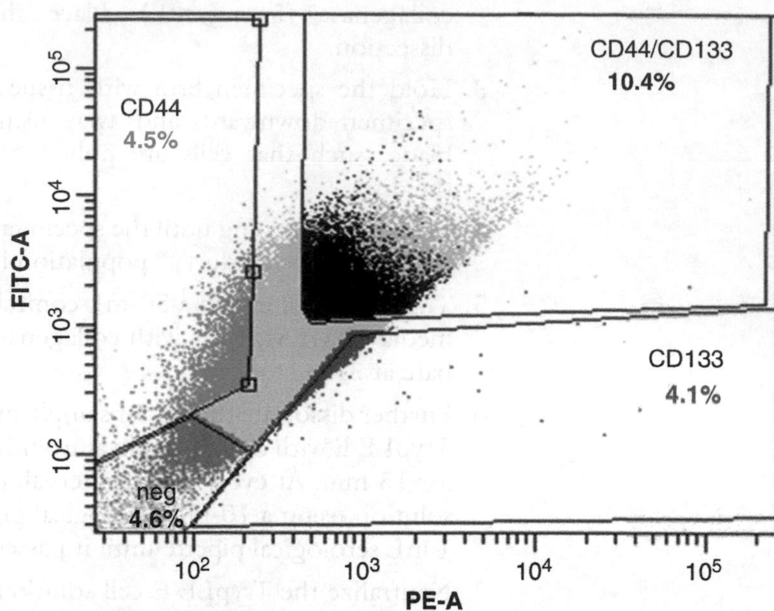

Fig. 1 Analysis of cancer stem cells through flow cytometry. FaDu cells were stained with CD44-FITC and CD133-PE and analyzed through flow cytometry

3.2.2 Hoechst Dye Exclusion: The Side Population

Side population by Hoechst dye exclusion technique helps in identification and sorting of stem cells in different tissues. Stem cells used to exclude Hoechst 33342 dye, which preferentially binds to DNA and the low fluorescence, can be detected by a flow cytometry.

1. Prepare single viable cell suspension from established cancer cell lines or from Subheading 3.1 as per the experimental setup (about 1×10^6 cells for analysis and 1×10^7 cells for sorting).
2. Resuspend 1×10^6 cells per mL in DMEM containing 0.2% FBS.
3. Add 5 μg/mL Hoechst33342 dye per 1×10^6 cells alone or with drug inhibitors like verapamil or reserpine and incubate for 90 min in dark at 37 °C. *see* **Note 7**.
4. Centrifuge the cells and resuspend it in HBSS containing 2 μg/mL propidium iodide to distinguish the dead cells.
5. Acquire the cells by flow cytometry. *See* **Note 8**.

3.2.3 Enzymatic Activity: ALDEFLUOR Assay

The ALDEFLUOR assay is a non-immunological method for identification of stem cells and progenitor cells based on their aldehyde dehydrogenase (ALDH) activity.

1. Prepare single viable cell suspension from established cancer cell lines or from Subheading 3.1 as per the experimental setup (about 1×10^6 cells for analysis and 1×10^7 cells for sorting).

2. Wash the cells with 1× sterile PBS as required. *See* **Note 9**.
3. Pellet down and suspend the cell pellet in ALDEFLUOR assay buffer containing ALDH substrate.
4. Add 5 μL of activated ALDEFLUOR reagent per 1×10^6 cells alone or with ALDEFLUOR™ DEAB and incubate for 40 min in dark at 37 °C. *See* **Note 10**.
5. Centrifuge the tubes and discard supernatant.
6. Resuspend the cell pellets in 500 μL of ALDEFLUOR™ Assay buffer and keep the suspension on ice or at 4 °C in dark.
7. Analyze the cells through flow cytometry.

3.2.4 Cell Culture Selection: Orosphere Assay

1. Prepare single viable cell suspension from established cancer cell lines or from Subheading 3.1 as per the experimental setup.
2. Seed 5000 cells/well in six-well low attachment plate with orosphere media for each experimental condition and place the cells in an incubator at 37 °C with 5% CO_2 for 7–10 days (Fig. 2). Do not disturb in between. *See* **Note 11**.
3. After 7–10 days of incubation (*see* **Note 12**), count the spheroid formed under a phase-contrast microscope using the 20× magnification lens.
4. Prepare single-cell suspension by dissociating orospheres with Accutase for 15 min at 37 °C followed by centrifugation and resuspend the cells in orosphere media.
5. For second and third generation of orospheres, equal number (5000 cells/well) of these live cells from the dissociated orospheres was reseeded as in **step 2**.

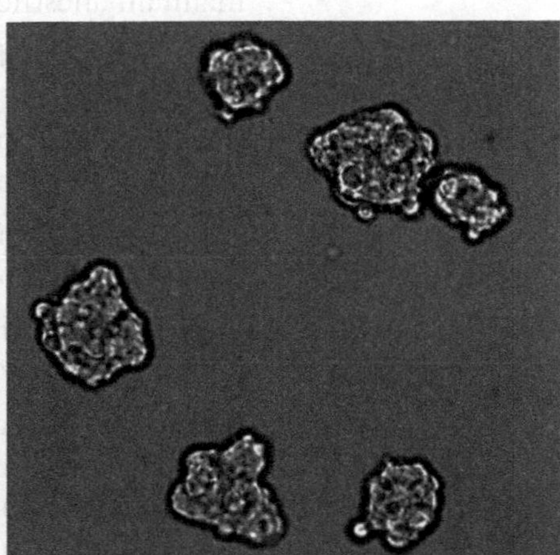

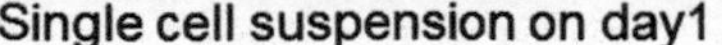

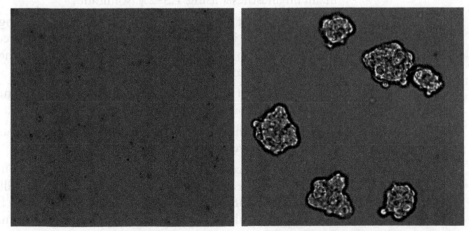

Single cell suspension on day1 Primary orosphere after 1 week

Fig. 2 Phase-contrast images of orospheres from FaDu cells. About 5000 cells/well were seeded in an orosphere media in an ultralow adherent six-well plate at 37 °C and after 7 days, the orospheres were observed in microscope, and photographed

3.3 Determining Tumorigenicity of Cancer Stem Cell by Xenograft Formation Assay

Apart from the above-discussed protocols used to isolate and characterize CSCs, xenograft formation assay is the gold standard in determining enhanced tumorigenicity of CSCs. As CSCs have increased tumorigenicity, they will be able to form tumors at a lower cell count as compared to a non-CSC population (*see* **Note 13**). By this method, exact amount of cells injected into NOD–SCID mice which results in tumor formation in 50% of the mice injected can be measured.

1. Collect CSC populations from previous protocols and place them on ice while preparing for injection.
2. Serially dilute the CSC-positive populations and the non-CSC populations with a wide range of cell densities (i.e., 1×10^6 cells; 2.5×10^5 cells; 1×10^5 cells; 2.5×10^4 cells; 5×10^3 cells; 1×10^3 cells; and finally 100 cells) in a serum-free media. So that the final concentration is equal to the desired number of cells per 100 μL.
3. Prepare enough volume of cell suspension at various concentrations in order to implant in at least four mice per dilution and cell population type in the subcutaneous tissue on the flank of the mouse. *See* **Note 14**.
4. Thaw the Matrigel by keeping it on ice, mix the cell suspension and Matrigel together at a ratio of 1:1 for injection, and draw up into a syringe with a 25-guage needle. *See* **Note 15**.
5. Place the mice into anesthesia induction chamber and induce anesthesia using 5% isoflurane at a rate of 0.7 L O_2/min.
6. Once the mice are unconscious and do not respond to a foot squeeze to test for pain, transfer the mice to a nose cone and maintain anesthesia using 1.5–2% isoflurane.
7. Clean the skin of the mice with a betadine scrub. If using hair-bearing mice, it is beneficial to have previously removed hair over the area to be injected, by either clipping or chemical methods. *See* **Note 16**.
8. Inject 200 μL of Matrigel–cancer cells mixture (1:1) into the mice subcutaneously. *See* **Note 17**.
9. Place the mice in a new cage using an appropriate warming method to recover from the anesthesia.
10. Observe the mice periodically for tumor formation. Observe the mice for at least 4 months as most of the tumors will form within 2–3 months, sometimes as quickly as 1 month before declaring a population as negative regarding tumor formation. *See* **Note 18**.

4 Notes

1. The decision to use a mechanical or a chemical-based dissociation is primarily a personal preference, but most often guided by the density of the tumor, site of collection, and patient heterogeneity. Some tumor types will be more amenable to mechanical dissociation with high viability, while others will be more appropriate for chemical digestion.
2. Use of the specimen cannot compromise patient care, so collaboration with the surgeon and pathologist who can quickly provide excess tissue is essential. Whereas, mouse tumors can be obtained more quickly, as they are resected immediately after sacrifice, can be identified grossly, and processing can begin quickly.
3. Clogging of filter (suspension with large debris) can be avoided by passing the suspension through 200 μm and 100 μm filters, respectively, and then proceed to 70 μm filters. Depending on the density of the tumor, eventually one may be able to aspirate the slurry into a 16-guage needle on a 3- or 5-mL syringe simply by this mechanical dissociation.
4. Washing should be done at room temperature for 10 min at 1000 rpm.
5. Directly add the antibody to the cell suspension as per the instruction given by the company.
6. Take an unstained cell mass as control.
7. Hoechst stain must be pre-warmed in a water bath just prior to the use rather than for a longer time interval.
8. Hoechst is a light-sensitive dye, so while handling the stained cells always avoid direct exposure to the light. Keep the stained cells over ice until the flow cytometer analysis.
9. The disrupted cells should be filtered through sterile 40 μm membranes to form single-cell suspensions.
10. Recap control tube and DEAB vial immediately to prevent evaporation as ALDEFLUOR™ DEAB is provided in 95% ethanol.
11. B27 supplement should be added freshly. Keep the cells on ice while not in use for the duration of the experiment. For different cell types, the seeding density varies. Do not disturb the plate.
12. The medium should not be changed or added to avoid disturbance of tumor spheres formation. To avoid evaporation of medium, seal the edges of the plate.
13. There is no established standard for how much more tumorigenic a population should be to be considered a CSC, but in

general, at least 50–100-fold less cells (compared to a marker-negative population) should be observed to achieve tumor formation in 50% of mice.

14. Tumor cells can be injected in different sites depending on the design of the experiment, the tumor cell type, and mouse strain. For example, if studying breast cancer, cells can be easily injected into the mammary fat pads of the mouse and still followed for development.
15. While handling Matrigel, keep all the necessary items (pipette, 200 μL tips, and syringe) on ice, as it will solidify at room temperature. Mixing cells with Matrigel is optional but likely improves tumor formation rate.
16. Xenograft formation is not solely dependent on the population—the choice of mouse strain can have an impact on the rate of tumor formation. NOD–SCID is an ideal model for injecting human tumor cells because of complete knockout of their immune system. Similarly, IL2 receptor-γ chain-deficient mice, which also lack NK cell function, require fewer cells to establish xenografts, but due to their severe immunocompromised state are difficult to maintain. Therefore, while choosing mouse model, it is important to weigh the benefits and risks of each model in terms of the study outcome.
17. For subcutaneous injection, use one hand to hold the skin of the mouse's flank taunt, insert the needle bevel up at a shallow angle (45° to the body) to prevent from excess penetration, i.e., beyond the subcutaneous layer. In case of proper insertion, a small bump should form on the skin after injecting cells.
18. In case of cancer cell line, the time for tumor formation will most likely be lower than when working with a primary patient tumor sample. Moreover, some cancers naturally grow faster than others do. However, it is important to have a fixed time for the endpoint of the experiment, as this will allow for a determination of amount of cells that leads to 50% tumor formation and an accurate determination if the cancer stem cell population is more tumorigenic than the non-cancer stem cell population.

Acknowledgments

Research support was partly provided by Board of Research in Nuclear Sciences (BRNS) [Number: 37(1)/14/38/2016-BRNS/37276], Department of Atomic Energy (DAE), and Science and Engineering Research Board (SERB) [Number: EMR/2016/001246]. Research infrastructure was partly provided by Fund for Improvement of S&T Infrastructure in Universities and Higher Educational Institutions (FIST) [Number: SR/FST/LSI-025/2014], Department of Science and Technology, Government of India.

References

1. Naik PP, Das DN, Panda PK, Mukhopadhyay S, Sinha N, Praharaj PP, Agarwal R, Bhutia SK (2016) Implications of cancer stem cells in developing therapeutic resistance in oral cancer. Oral Oncol 62:122–135
2. Sinha N, Mukhopadhyay S, Das DN, Panda PK, Bhutia SK (2013) Relevance of cancer initiating/stem cells in carcinogenesis and therapy resistance in oral cancer. Oral Oncol 49(9):854–862
3. Johnson S, Chen H, Lo P-K (2013) *In vitro* tumorsphere formation assays. Bio-protocol 3 (3):e325
4. Greve B, Kelsch R, Spaniol K, Eich HT, Götte M (2012) Flow cytometry in cancer stem cell analysis and separation. Cytometry A 81 (4):284–293
5. Ginestier C, Hur MH, Charafe-Jauffret E, Monville F, Dutcher J, Brown M, Jacquemier J, Viens P, Kleer CG, Liu S, Schott A, Hayes D, Birnbaum D, Wicha MS, Dontu G (2007) ALDH1 is a marker of normal and malignant human mammary stem cells and a predictor of poor clinical outcome. Cell Stem Cell 1 (5):555–567

Methods in Molecular Biology (2019) 2002: 141–150
DOI 10.1007/7651_2018_178

Published online: 12 August 2018

Determining Competitive Potential of Bone Metastatic Cancer Cells in the Murine Hematopoietic Stem Cell Niche

Sun H. Park, Matthew R. Eber, Russell S. Taichman, and Yusuke Shiozawa

Abstract

The ability of cancer cells to compete with hematopoietic stem cells (HSCs) to target the bone marrow microenvironment, or the HSC niche, during the dissemination process is critical for the development of bone metastasis. Here, we describe the methods for testing the relative potential of cancer cells to compete with HSCs for occupancy of the HSC niche by measuring the peripheral blood level of engrafted HSCs by flow cytometry in mice after bone marrow transplantation and tandem cancer cell inoculation. This method is useful for determining the molecular mechanisms for the roles of HSCs in the regulation of bone metastases.

Keywords Bone homing, Bone marrow transplantation, Disseminated tumor cells, Flow cytometry, Hematopoietic stem cell, HSC niche, Mouse model, Niche competition

1 Introduction

Hematopoietic stem cells (HSCs) give rise to mature blood cells (e.g., myeloid and lymphoid lineages) during a process known as hematopoiesis. Besides differentiation potential, to maintain the homeostasis of hematopoiesis HSCs must be capable of self-renewal, similar to other stem cell types. Although most HSCs reside within the bone marrow, it is known that a portion of HSCs circulate throughout the peripheral blood. These circulating HSCs eventually home to the bone marrow [1]. It has been suggested that the specific microenvironment in the marrow, or the "niche," is an indispensable component for controlling HSC differentiation, self-renewal, and homing [2]. Growing evidence suggests that osteoblasts and their progenitor cells are an important element of the HSC niche, although bone marrow stromal cells (endothelial cells, macrophages, fibroblasts, pericytes, and adipocytes) are also known to serve as HSC niche components [3, 4]. Indeed, chemotactic factors released from osteoblasts and

Sun H. Park and Matthew R. Eber contributed equally to this work.

other niche components [e.g., C-X-C motif chemokine 12 (CXCL12), and stem cell factor (SCF)] help to regulate essential HSC functions through the receptors expressed on HSCs [e.g., C-X-C chemokine receptor type 4 (CXCR4), c-Kit] [2, 5–7].

Interestingly, it has recently been revealed that disseminated tumor cells (DTCs) use bone homing mechanisms similar to HSCs to gain access to the marrow [8]. Moreover, once in the marrow, DTCs directly compete with HSCs to occupy the HSC niche and survive within the bone microenvironment [8]. The HSC niche also induces DTCs to undergo cellular dormancy, a state similar to HSC quiescence. It has been suggested that these dormant DTCs acquire treatment resistance and escape from immune surveillance, eventually developing into full-blown bone metastases [9, 10]. Therefore, the HSC niche is a critical player in bone metastatic progression. However, little is known as to the molecular mechanisms whereby DTCs interact with HSCs to establish bone metastases.

In this protocol, we describe the methods necessary to determine the potential of cancer cells to compete with HSCs for the bone marrow HSC niche by combining cancer cell inoculation techniques with bone marrow transplantation. The workflow of these methods to address specific questions in HSC niche competition is outlined in Fig. 1. These models can be further utilized to understand the mechanisms that control bone marrow stem cell

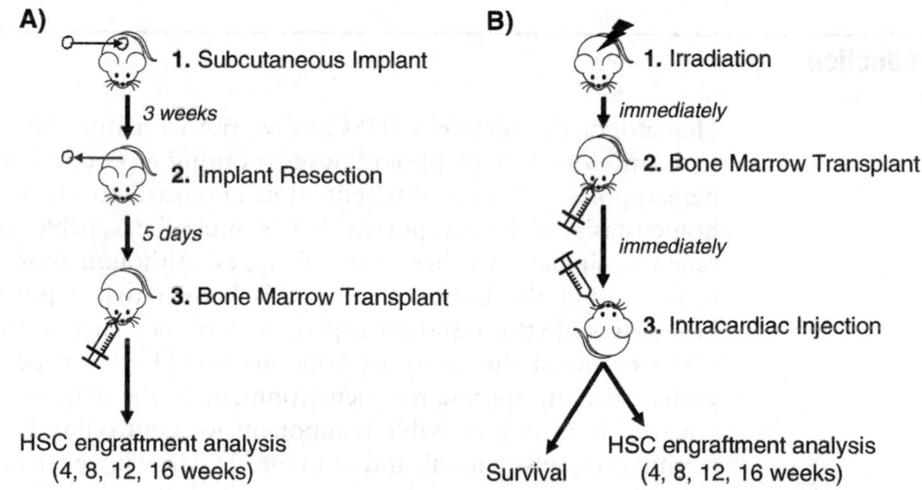

Fig. 1 Hematopoietic stem cell (HSC) niche competition assay workflow. (**a**) In the first model, a cancer cell line is implanted subcutaneously, grown for 3 weeks and then resected. Five days later, a bone marrow transplant is performed. (**b**) In the second model, a bone marrow transplant is performed after total body irradiation, and then a cancer cell line is inoculated systemically with intracardiac injection. In both models, the readout for HSC niche competition is HSC engraftment analysis by flow cytometry of peripheral blood. Additionally, a survival outcome can be used in the second model

niche homing of highly competitive cancer cells in order to promote the development of novel therapeutic strategies for bone metastasis.

2 Materials

2.1 Bone Marrow Transplant

1. Five- to Eight-week-old mice whose leukocytes express the cluster of differentiation (CD)45.2 or Ly5.2, such as C57BL/6J (Jackson Laboratory)
2. Five- to Eight-week-old mice whose leukocytes express CD45.1 or Ly5.1, such as B6.SJL-*Ptprc*^{a} *Pepc*^{b}/BoyJ (Jackson Laboratory)
3. CO_2 euthanasia chamber
4. Biosafety cabinet
5. 70% Ethanol solution
6. Sterile surgical instruments (scissors, scalpel, and forceps)
7. 3 mL Syringe with 25-g needle (Becton Dickinson)
8. Dulbecco's phosphate-buffered saline without calcium or magnesium (DPBS): 20 mg/L KCl, 200 mg/L KH_2PO_4, 8.00 g/L NaCl, 2.16 g/L Na_2HPO_4, and pH 7.2–7.4
9. Fetal bovine serum (FBS)
10. Fluorescence activated cell sorting (FACS) buffer: 5% FBS, DPBS
11. 10-cm Tissue culture (TC) dish
12. 40-μm Mesh strainer
13. 50-mL Conical tube
14. Hemocytometer and trypan blue solution, 0.4%
15. 0.9% Sterile saline solution
16. Isoflurane vaporizer
17. A gamma cell 40 cesium source
18. 28-g Insulin syringe

2.2 Cancer Cell Inoculation

1. Cancer cell of interest
2. DPBS
3. Biosafety cabinet
4. 40-μm Mesh strainer
5. 50-mL Conical tube
6. Hemocytometer and trypan blue solution, 0.4%
7. 1.5-mL Microcentrifuge tubes
8. 0.9% Sterile saline solution

9. Five- to Eight-week-old mice
10. Electrical clippers or a depilatory agent
11. Isoflurane vaporizer
12. Povidone-iodine, 5%
13. Sterile alcohol prep pads
14. 28-g Insulin syringe
15. Gelfoam® gelatin sponges (Pfizer)
16. 10-cm TC plate
17. Sterile surgical gauze
18. Sterile surgical instruments (scissors and forceps)
19. Surgical wound clips
20. Heating pad

2.3 HSC Engraftment Analysis

1. Heating pad
2. Sterile razor blades
3. 5-mL Round bottom tubes
4. Peripheral blood collection buffer: 5 μM ethylenediaminetetraacetic acid (EDTA), DPBS
5. Red blood cell (RBC) lysis buffer (BioLegend)
6. DPBS
7. FACS buffer
8. Anti-CD45.1 and CD45.2 antibodies (BioLegend)
9. Flow cytometer

3 Methods

3.1 Bone Marrow Transplant

1. Euthanize and douse the skin of the donor mice in 70% ethanol (*see* **Note 1**). Remove the hind limbs and store in a 10-cm TC dish containing ice cold sterile DPBS on ice.
2. In a biosafety cabinet, separate the femur from the tibia and remove the excess tissue thoroughly using a scalpel. Be careful to maintain the integrity of the bone in order to maintain sterility of the bone marrow. Store the bare bones in a 10-cm TC dish containing ice cold sterile DPBS.
3. Remove the epiphyses from the femurs and tibias using a scalpel.
4. Flush the marrow from the diaphysis into a new sterile 10-cm TC dish with ice cold sterile FACS buffer using a 25-g needle and 3-mL syringe.

5. Pass the marrow several times through the syringe to dissociate the marrow pieces, and then pass the marrow suspension through a 40-μm mesh strainer into a 50-ml conical tube.
6. Determine the bone marrow cell number using a hemocytometer and the trypan blue exclusion assay and then pellet the bone marrow cells by centrifugation (350 × *g* for 5 min).
7. Aspirate the supernatant and resuspend the bone marrow cells in ice cold sterile 0.9% saline at a concentration of 2×10^5 cells/100 μL and store on ice (*see* **Note 2**).
8. If the intracardiac injection model is to be used, the recipient mice must be irradiated prior to bone marrow transplantation. Radiation dose is animal model dependent (*see* **Note 3**).
9. Anesthetize the recipient mouse and perform retro-orbital sinus injection of 2×10^5 bone marrow cells/100 μL per each mouse.
10. Allow the animal to recover from anesthesia in a clean cage placed on a heating pad.
11. The engraftment of donor cells will be followed using flow cytometry as described in Subheading 3.3.

3.2 Cancer Cell Inoculation

1. Wash and harvest the cells to be tested under sterile conditions (*see* **Note 4**).
2. Pass the cells through a 40-μm cell strainer into a 50-mL conical tube and determine cell number as before with a hemocytometer and the trypan blue exclusion assay and then pellet the cells by centrifugation (180 × *g* for 5 min).
3. Aspirate the supernatant and resuspend the cells in ice cold sterile 0.9% saline to an appropriate concentration and volume and transfer to a sterile microcentrifuge tube and store on ice (*see* **Note 5**).
4. If performing subcutaneous implant of the inoculum, cut the gelatin foam into 3 mm^3 squares (scaffolds) using surgical scissors and with forceps place them in a 10-cm TC dish containing ice cold 0.9% saline. This must be performed under sterile conditions.
5. Anesthetize the recipient mouse using isoflurane inhalation and prepare it for the surgery to be performed, either subcutaneous implant and resection (Subheading 3.2.1) or intracardiac injection (Subheading 3.2.2) (*see* **Note 6**).

3.2.1 Subcutaneous Implant and Resection

1. If the recipient mouse has fur on its back, it must first be removed using electrical clippers or a depilatory agent.

2. Place the anesthetized mouse in the prone position and disinfect the skin of the back using a povidone-iodine scrub and alcohol prep pad.
3. Lift up a tent of skin midway up the back using forceps and make a 1-cm horizontal incision using blunt-tipped surgical scissors.
4. Use blunt-tipped surgical scissors to create a 1–2-cm subcutaneous pouch rostral to the incision site.
5. Blot a 3-mm^3 gelatin scaffold on sterile surgical gauze to remove excess liquid and then saturate it with inoculum and place it deep into the subcutaneous pouch using forceps.
6. Release the implant and withdraw the forceps without disturbing the placement of the implant.
7. Close the incision using a surgical wound clip.
8. Allow the animal to recover from anesthesia in a clean cage placed on a heating pad.
9. Remove the wound clip after 7 days or when the incision has healed.
10. Three weeks after implantation, anesthetize the mouse using isoflurane inhalation. If the animal has fur on its back, it must first be removed using electrical clippers or a depilatory agent.
11. Place the anesthetized mouse in the prone position and disinfect the skin of the back using a povidone-iodine scrub and alcohol prep pad.
12. Make an incision near the subcutaneous implant with blunt-tipped surgical scissors.
13. Lift up a tent of skin near the incision/implant and use blunt-tipped surgical scissors to reopen the subcutaneous pouch.
14. Dissect the implant away from the skin and surrounding tissue and remove it entirely.
15. Close the incision using a surgical wound clip.
16. Allow the animal to recover from anesthesia in a clean cage placed on a heating pad.
17. Remove the wound clip after 7 days or when the incision has healed.

3.2.2 Intracardiac Injection

1. If the animal has fur on its chest, it must first be removed using electrical clippers or a depilatory agent.
2. Place the anesthetized mouse in the supine position and disinfect the skin of the chest using a povidone-iodine scrub and alcohol prep pad.
3. Load 100 μL of inoculum into an insulin syringe.

4. Use the index and middle finger of your nondominant hand to secure the limbs of the mouse (*see* **Note 7**).
5. Using your dominant hand, insert the needle perpendicular about 7–8 mm into the injection point, 1–2 mm left of the midline halfway between the clavicle and xiphoid process.
6. Using your nondominant hand, pull up very slightly on the plunger being careful not to move the needle. If the needle is in the correct position (left ventricle), you will see a pulse of bright red blood flush into the inoculum.
7. Inject the inoculum slowly over about 30 s. During the injection, periodically confirm that the needle is still in the left ventricle by pulling up on the plunger slightly to check for the signature pulse of bright red blood (*see* **Note 8**).
8. Gently remove the needle.
9. Allow the animal to recover from anesthesia in a clean cage placed on a heating pad.

3.3 HSC Engraftment Analysis

1. Place the cage of animals to be analyzed on the heating pad and allow it to warm up for 10 min.
2. Place a mouse on top of the wire food rack and thread its tail through to the bottom.
3. Make a small nick in the tail vein perpendicular to the tail with a razor blade and collect the blood droplets in a 5-mL round bottom tube containing 4-mL peripheral blood collection buffer (*see* **Note 9**). Use the lip of the tube to catch the blood droplets and be sure that they mix with the collection buffer by tapping the tube gently in order to prevent coagulation.
4. Apply pressure to the wound with a piece of gauze and confirm hemostasis of the tail wound before placing the mouse back in its cage.
5. Centrifuge the collection tubes (350 × *g* for 5 min).
6. Decant the supernatant and resuspend the pellets in 500 μL RBC lysis buffer and incubate at room temperature for 5 min (*see* **Note 10**).
7. Fill each tube with 4 mL DPBS and then centrifuge them (350 × *g* for 5 min).
8. Decant the supernatant and resuspend the pellets in 500 μL FACS buffer containing anti-CD45.1 and CD45.2 antibodies (*see* **Note 11**).
9. Incubate on ice in the dark for 1 h.
10. Fill each tube with 4 mL DPBS and then centrifuge them (350 × *g* for 5 min).

11. Decant the supernatant and resuspend the pellets in 500 μL FACS buffer.
12. Analyze the samples for expression of host hematopoietic cells versus expression of donor hematopoietic cells by distinguishing CD45.1 and CD45.2 positive cells by flow cytometry (*see* **Note 12**).

4 Notes

1. The donor mice must express a different CD45 allele than the recipient mice in order to determine HSC engraftment. We recommend performing the HSC engraftment analysis (Subheading 3.3) on representative naïve donor and recipient mice before transplantation to verify distinguishability by flow cytometry.
2. Instead of mouse donor cells, it is possible to use human HSCs at the same concentration, as long as the recipient animal is immunocompromised. In this case, we would recommend using NOD.Cg-Prkdc^{scid} Il2rg^{tm1Wjl}/SzJ (NSG) mice as a recipient.
3. If greater HSC engraftment is desired, the recipient mice can be irradiated immediately before transplantation. For lethal irradiation levels in immunocompetent mice, we recommend a total body dose of 1140 cGy (two 570 cGy doses, 3 h apart). For lethal irradiation levels in immunocompromised mice, we recommend a total body dose of 700 cGy (two 350 cGy doses, 3 h apart). If human HSC transplant is to be performed in immunocompromised mice, we recommend a sublethal total body dose of 300 cGy (two 150 cGy doses, 3 h apart).
4. In order to remove the variable of implanted cell growth after intracardiac injection, it is possible to irradiate the cell line to be tested immediately before harvest and implantation. In our studies using prostate cancer cells, we first irradiated them with 8 Gy (two 4 Gy doses, 3 h apart). However, irradiation dose will be cell dependent. In this case, a dose that causes cell growth arrest but not apoptosis should be used.
5. The concentration of the cell suspension depends on the aggressiveness of the cell line and the inoculation to be performed. As a general rule, the subcutaneous implant will require 10 μL of inoculum whereas the intracardiac injection requires 100 μL. In our studies using prostate cancer cells, we used 2×10^5 cells/10 μL for the subcutaneous implant and 2×10^6 cells/100 μL for the intracardiac injection. Very aggressive cell lines may not require as many cells, but these are good starting points. For ease of use, we recommend

creating individual tubes containing the inoculum required for each subcutaneous implant.

6. The model chosen depends on the hypothesis to be tested. The subcutaneous implant model relies on the ability of the cell line to disseminate from the primary implant to the bone on its own. While this closely models what happens clinically, the number of disseminated cells in the bone is often quite low and is entirely cell line dependent. The intracardiac model floods the circulation with many cells which increases "dissemination" to the bone, with the caveat that it does not accurately model the dissemination process. Also, using the intracardiac model it is possible to employ an engraftment readout using animal survival if the recipient animals are first lethally irradiated. In this model, longer survival times signify greater HSC engraftment.
7. It is essential that the mouse is flat on its back with its internal organs centered. It is common for the mouse to shift its weight naturally to favor one side over the other, but if this happens it is difficult to reliably locate the left ventricle with the needle. When the mouse is in the correct position, use your hand to secure it in place without using excessive pressure. Pressure on the abdomen causes shifting of the internal organs, again making the injection more challenging.
8. If the action of pulling up on the plunger creates a vacuum and not a pulse of blood, the needle has either drifted too far up or too deep and is no longer in the left ventricle. If this happens, slightly push the needle deeper or withdraw it until the pulse of blood is observed before proceeding with the injection.
9. Create a nick about halfway down the tail so that if the initial cut fails to produce a blood droplet it is possible to make another nick rostral to the initial cut. Do not squeeze or attempt to milk the tail, as this can cause contamination with lymphatic fluid. In our experience, five drops (~10–50 μL) is sufficient for analysis.
10. This process can be repeated until the pellet is no longer red, or can be skipped entirely if there is concern for lost cells during RBC lysis.
11. It is recommended to use antibodies conjugated to fluorophores that have emission spectrums that do not overlap much, such as fluorescein isothiocyanate (FITC) and allophycocyanin (APC). Concentrations and multiplexing of antibodies require in lab optimization.
12. It is useful to perform HSC engraftment analysis many times over a period of several weeks. We have found that hematopoietic cells of transplanted origin can be detected up to 16 weeks after transplant.

Acknowledgments

This work is directly supported by Department of Defense (W81XWH-14-1-0403, Y. Shiozawa; W81XWH-17-1-0541, Y. Shiozawa), the Wake Forest Baptist Comprehensive Cancer Center Internal Pilot Funding (Y. Shiozawa), and the Wake Forest School of Medicine Internal Pilot Funding (Y. Shiozawa). Y Shiozawa is supported by the Translational Research Academy which is supported by the National Center for Advancing Translational Sciences (NCATS), National Institutes of Health, through Grant Award Number UL1TR001420. This work is also supported by the National Cancer Institute's Cancer Center Support Grant award number P30CA012197 issued to the Wake Forest Baptist Comprehensive Cancer Center. The content is solely the responsibility of the authors and does not necessarily represent the official views of the National Cancer Institute. Sun H. Park and Matthew R. Eber have contributed equally to this work.

References

1. Hardy CL (1995) The homing of hematopoietic stem cells to the bone marrow. Am J Med Sci 309(5):260–266
2. Taichman RS (2005) Blood and bone: two tissues whose fates are intertwined to create the hematopoietic stem-cell niche. Blood 105 (7):2631–2639. https://doi.org/10.1182/blood-2004-06-2480
3. Wilson A, Trumpp A (2006) Bone-marrow haematopoietic-stem-cell niches. Nat Rev Immunol 6(2):93–106. https://doi.org/10.1038/nri1779
4. Yin T, Li L (2006) The stem cell niches in bone. J Clin Invest 116(5):1195–1201. https://doi.org/10.1172/JCI28568
5. Aiuti A, Tavian M, Cipponi A, Ficara F, Zappone E, Hoxie J, Peault B, Bordignon C (1999) Expression of CXCR4, the receptor for stromal cell-derived factor-1 on fetal and adult human lympho-hematopoietic progenitors. Eur J Immunol 29(6):1823–1831. https://doi.org/10.1002/(SICI)1521-4141(199906)29:06<1823::AID-IMMU1823>3.0.CO;2-B
6. Nagasawa T, Hirota S, Tachibana K, Takakura N, Nishikawa S, Kitamura Y, Yoshida N, Kikutani H, Kishimoto T (1996) Defects of B-cell lymphopoiesis and bone-marrow myelopoiesis in mice lacking the CXC chemokine PBSF/SDF-1. Nature 382 (6592):635–638. https://doi.org/10.1038/382635a0
7. Peled A, Petit I, Kollet O, Magid M, Ponomaryov T, Byk T, Nagler A, Ben-Hur H, Many A, Shultz L, Lider O, Alon R, Zipori D, Lapidot T (1999) Dependence of human stem cell engraftment and repopulation of NOD/SCID mice on CXCR4. Science 283 (5403):845–848
8. Shiozawa Y, Pedersen EA, Havens AM, Jung Y, Mishra A, Joseph J, Kim JK, Patel LR, Ying C, Ziegler AM, Pienta MJ, Song J, Wang J, Loberg RD, Krebsbach PH, Pienta KJ, Taichman RS (2011) Human prostate cancer metastases target the hematopoietic stem cell niche to establish footholds in mouse bone marrow. J Clin Invest 121(4):1298–1312. https://doi.org/10.1172/JCI43414
9. Baxevanis CN, Perez SA (2015) Cancer dormancy: a regulatory role for endogenous immunity in establishing and maintaining the tumor dormant state. Vaccines (Basel) 3 (3):597–619. https://doi.org/10.3390/vaccines3030597
10. Decker AM, Jung Y, Cackowski F, Taichman RS (2016) The role of hematopoietic stem cell niche in prostate cancer bone metastasis. J Bone Oncol 5(3):117–120. https://doi.org/10.1016/j.jbo.2016.02.005

Methods in Molecular Biology (2019) 2002: 151–163
DOI 10.1007/7651_2018_188

Published online: 08 September 2018

Spatial Genomic Analysis: A Multiplexed Transcriptional Profiling Method that Reveals Subpopulations of Cells Within Intact Tissues

Antti Lignell and Laura Kerosuo

Abstract

Here, we present Spatial Genomic Analysis (SGA), a quantitative single-cell transcriptional profiling method that takes advantage of single-molecule imaging of individual transcripts for up to a hundred genes. SGA relies on a machine learning-based image analysis pipeline that performs cell segmentation and transcript counting in a robust way. SGA is suitable for various *in situ* applications and was originally developed to address heterogeneity in the neural crest, which is a transient embryonic stem cell population important for formation of various vertebrate body structures. After being specified as multipotent neural crest stem cells in the dorsal neural tube, they go through an epithelial to mesenchymal transition in order to migrate to different destinations around the body, and gradually turn from stem cells to progenitors prior to final commitment. The molecular details of this process remain largely unknown, and upon their emergence, the neural crest cells have been considered as a single homogeneous population. Technical limitations have restricted the possibility to parse the neural crest cell pool into subgroups according to multiplex gene expression properties. By using SGA, we were able to identify subgroups inside the neural crest niche in the dorsal neural tube. The high sensitivity of the method allows detection of low expression levels and we were able to determine factors not previously shown to be present in neural crest stem cells, such as pluripotency or lineage markers. Finally, SGA analysis also provides prediction of gene relationships within individual cells, and thus has broad utility for powerful transcriptome analyses in original biological contexts.

Keywords Chicken embryo, HCR, Hybridization chain reaction, In vivo single-cell analysis, Neural crest stem cell niche, Neural crest stem cells, Pluripotency, Single-molecule microscopy, Quantitative single-molecule fluorescent in situ hybridization, SGA, smFISH, Spatial genomic analysis, Spatial genomics, Spatial tissue transcriptome analysis

1 Introduction

Recent development in single-cell transcriptional profiling largely relies on single-cell RNA sequencing-based methods [1]. Despite the transcriptome-level throughput of these applications, they lack spatial orientation of cells in host tissues. On the other hand, single-molecule fluorescent *in situ* hybridization (smFISH) [2] is a useful method for quantitative analysis of individual transcripts in single cells, but low signal and small throughput has so far limited its use mainly for cell culture approaches with simultaneous detection of only a handful of genes. We have established a Spatial Genomic Analysis (SGA) pipeline [3] that overcomes these limitations by using hybridization chain reaction (HCR) signal amplification [4, 5] coupled with sequential hybridization rounds [6], as shown in Fig. 1. In order to achieve single-cell resolution in complex tissue samples, SGA contains a machine learning algorithm-based cell segmentation as well as a dot counting routine of the individual transcripts [3]. Cells are divided into subgroups by using unbiased hierarchical clustering based on their multiplexed gene expression profile, and then mapped back to their original spatial context in the tissue. SGA is a powerful method that can be applied to identify transcriptionally distinct subpopulations in stem cell niches or other heterogeneous cell populations according to expression profiles of tens or even up to a hundred genes.

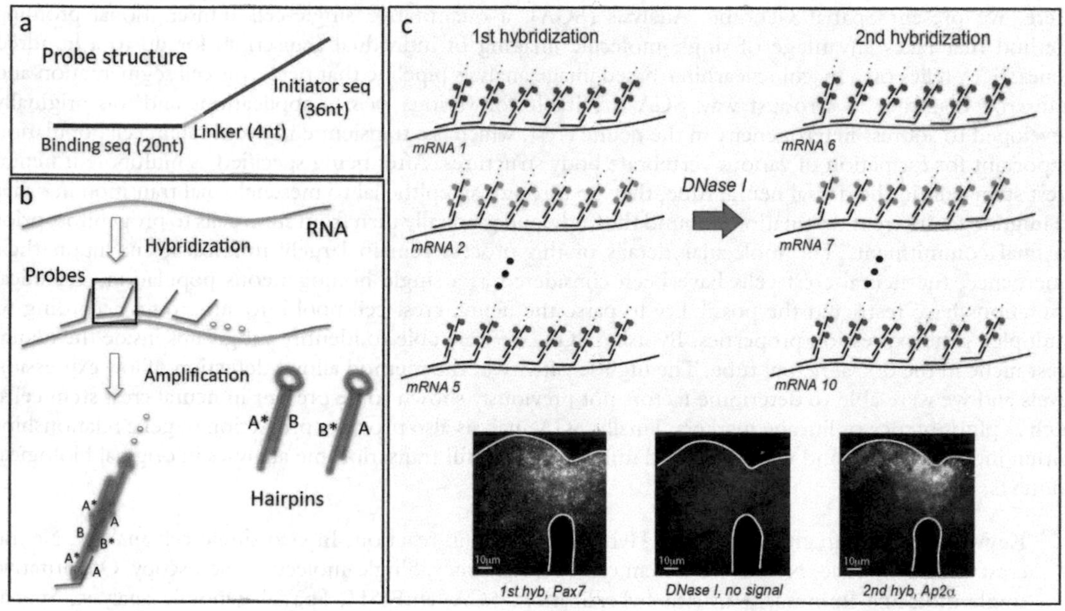

Fig. 1 Signal amplification by using hybridization chain reaction and the sequential hybridization steps in SGA. (**a**) Each SGA probe consists of a 20-nucleotide-long CDS recognition sequence followed by a four-nucleotide linker sequence before the initiator sequence in the 3′ end required for the HCR amplification. (**b**) Each transcript is hybridized with up to 24 individual DNA probes tagged with the same initiator sequence. CDS of the mRNA is marked with green color. A pair of metastable hairpins tagged with fluorophores is used to amplify the signal for each individual probe. The chain reaction that leads to an amplified signal begins as the single-stranded initiator sequence tagged to the hybridized probe opens one hairpin, hybridizes into it, and then opens a second hairpin, which again reveals the original initiator sequence and so forth. (**c**) A cartoon of the multiplexing scheme where a set of five genes are hybridized and imaged followed by probe stripping with DNase I enzyme. This routine is repeated until the desired number of genes is reached. (Figure **c** is reprinted from Lignell A, Kerosuo L, Streichan SJ, Cai L, Bronner ME (2017) Identification of a neural crest stem cell niche by Spatial Genomic Analysis. Nature Communications 8(1):1830. doi: https:/doi.org/10.1038/s41467-017-01561-w with permission from Springer Nature)

We used SGA to address heterogeneity in the developing dorsal neural tube of chicken embryos where the neural crest cells are known to develop. Unbiased hierarchical clustering revealed five subpopulations with spatially distinct localization (Fig. 2a, b) providing novel insight into the mechanisms of neural crest development in its transient stem cell niche. Additionally, SGA allowed us to determine a core set of neural crest genes and draw conclusions on gene relationships based on their synexpression patterns within individual cells (Fig. 2c). Here, we describe the SGA method in full detail and provide suggestions for usage in other tissues.

2 Materials

Since RNA integrity is critical for the SGA method, all the buffers and chemicals should be purchased in RNase-free quality (when available) and with highest purity. The glassware, tubes, bench space, and gloves should be made RNase free and filtered pipette tips should be used. Appropriate personal protective equipment (safety goggles, lab jacket, nitrile gloves, and fume hood) should be used when working with hazardous and toxic chemicals. Use RNase-free water for all solutions.

2.1 Buffers

1. Standard buffers: 10× PBS and 20× SSC buffers should be purchased directly in RNase-free quality and aliquoted to 50 ml conical tubes and stored at room temperature. The working solutions should be made fresh from these stocks right before use.
2. Fixation solution: 4% paraformaldehyde in 1× PBS. The solution can be stored at −20 °C.
3. Anti-bleach buffer (ABB): The base (ABBB) is 20 nM tris–HCl, and 50 nM NaCl saturated with Trolox. This can be made in 50 ml conical tube by adding 120 mg tris–HCl, 146 mg NaCl, and 200 mg of Trolox into RNase-free water and vortexed thoroughly (*see* **Note 1**). All the components can be made in advance and stored in aliquots at −20 °C. First, make an 8% D-glucose solution in water and aliquot to microcentrifuge tubes. Second, make a pyranose oxidase solution by diluting the solid protein in its container with water and measure the concentration with a spectrometer until OD (405 nm) is 0.5, and then aliquot to microcentrifuge tubes. Finally, dilute the catalase enzyme into a 10-mg/ml concentration in water and aliquot to microcentrifuge tubes. All three components should be thawed at room temperature before use.
4. Incubation/hybridization buffers: Use high-molecular weight (>500,000) dextran sulfate (DS) to make a 10% solution (100 mg of DS mixed with 100 μl 20× SSC, 0–300 μl of formamide (FA) for 0%, 10%, and 30% solutions, respectively,

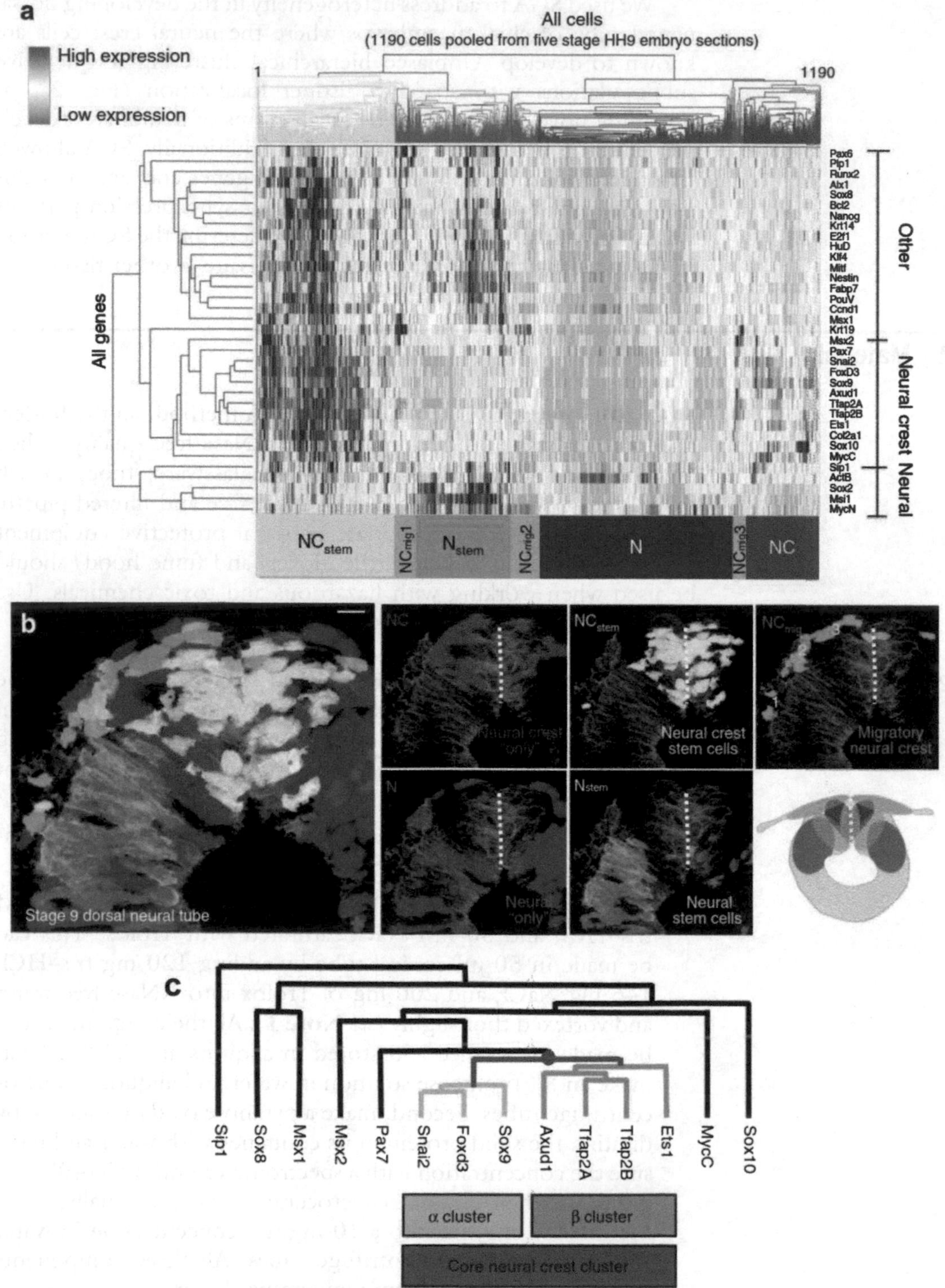

Fig. 2 (**a**) Unbiased hierarchical clustering analysis of 1190 cells in the cranial dorsal neural tube of an HH Stage 9 chicken embryo reveals distinct subpopulations according to transcriptional similarity of the 35 genes analyzed by using SGA. The NCstem (yellow) and NC (red) cells refer to premigratory neural crest cells with or

filled to 1 ml total volume with water in microcentrifuge tubes). Hybridization buffers that are used for overnight incubations should contain 0.02% thiomersal to prevent fungal and bacterial growth. These solutions should be made right before use.

5. Pre-hybridization buffers: Make 1% PFA in 1× PBS buffer from a PFA stock and store aliquots at −20 °C. Additionally, make 0.5% SDS in 1× PBS, and 1% $NaBH_4$ solution in 1× PBS (weigh 10 mg of $NaBH_4$ and mix with 1 ml of 1× PBS on a weight boat right before use, *see* **Note 2** for safety instructions).
6. Hairpin solutions: The hairpins labeled with fluorescent dye can be ordered from Molecular Technologies (www.moleculartechnologies.org), please find a detailed description of their structure in this reference [5].
7. Immunostaining and blocking buffers: Immunostaining is performed in 1× PBS/0.2% triton (PBT) with 5% bovine serum albumin (BSA) as a blocking and 1% DMSO as a permeabilization agent.

2.2 Cover Glasses

1. Treat #1.5 cover glasses with (3-aminopropyl)triethoxysilane (AS) to provide strong sample adherence to the glass surface as follows: first, sonicate cover glasses in a container filled with acetone for 1 h in an ultrasound bath, and then dip them into 2% AS solution in acetone followed by two additional washes with acetone. Finally, rapidly air-dry the cover glasses under compressed air (*see* **Note 3**).
2. After collecting the samples as cryosections on the cover glasses (*see* sample preparation in Subheading 3.1), cover the cryosections with hybridization chambers (e.g., 8 mm in diameter and 50 μl volume) that stick to the glass around the sample. The chambers maintain the operational volume small to assure efficient usage of the expensive reagents, and also prevent the

Fig. 2 (continued) without expression of pluripotency factors, respectively, and the green cells represent newly migrating neural crest cells (NCmig). Cells destined to become parts of the central nervous system are also divided into stem cell (Nstem) and "neural only" (N) subgroups. (**b**) Spatial orientation of the clustered cells in the neural tube reveals that the pluripotent neural crest stem cells are located around the midline. The colors refer to the cell clusters shown in the heatmap above. (**c**) Synexpression analysis showing the similarity of gene expression patterns in individual cells. A "core neural crest" cluster of genes is revealed showing the set of neural crest genes that are most likely to be expressed together in individual cells at this developmental time point, and which is further divided into two separate subgroups. This approach may provide a valuable addition for studies on relationships between genes in gene regulatory circuitries. (The figures **a–c** are reprinted from Lignell A, Kerosuo L, Streichan SJ, Cai L, Bronner ME (2017) Identification of a neural crest stem cell niche by Spatial Genomic Analysis. Nature Communications 8(1):1830. doi: https:/doi.org/10.1038/s41467-017-01561-w with permission from Springer Nature)

samples from drying and protect them from detaching. Imaging is performed from the bottom through the cover glass with an inverted microscope.

2.3 Probe Design

1. Coding sequences (CDS) of each mRNA can be acquired from your database of choice (e.g., www.ncbi.nlm.nih.gov) and, if possible, a probe set of up to 24 single-stranded DNA probes per CDS should be designed (the minimum number of probes per CDS we have successfully used is 13). Blast the sequences to ensure unique binding. Each probe consists of a 20-nucleotide reverse compliment DNA consensus sequence. Importantly, leave a separation gap of at least 5 nucleotides between the individual probes. GC content of the probes should be between 40 and 60% and melting temperatures around 52–58 °C. If the CDS is short, 18- or 19-nucleotide-long probe sequences can be used, while the melting temperature is maintained at the same range.
2. In addition to the CDS-binding sequence, a four-nucleotide linker sequence as well as a (B1–B5) HCR initiator sequence [5] is added to the 3′ end of each probe, resulting in a 60-nucleotide-long probe (Fig. 1a). The sequences are listed below, and the initiators are shown in capital and the linkers in lowercase letters. The HCR amplification scheme is presented in Fig. 1b.

   B1: tataGCATTCTTTCTTGAGGAGGGCAGCAAACGGGAAGAG

   B2: aaaaAGCTCAGTCCATCCTCGTAAATCCTCATCAATCATC

   B3: taaaAAAGTCTAATCCGTCCCTGCCTCTATATCTCCACTC

   B4: atttCACATTTACAGACCTCAACCTACCTCCAACTCTCAC

   B5: atttCACTTCATATCACTCACTCCCAATCTCTATCTACCC

   The probes can be ordered from an oligo manufacturer (e.g., www.idtdna.com) in a 96-well plate diluted to a 100-μM concentration. The probe sets for each gene are then combined and diluted into a 100-nM per probe concentration (100×) that is directly used for experiments. Both the 100-μM probe stocks and the 100-nM/probe working solutions are stored at −20 °C.

3 Methods

All the steps are performed at room temperature unless otherwise noted.

3.1 Sample Preparation and Cryosectioning

1. Fix samples overnight at 4 °C with 4% paraformaldehyde in 1× PBS. After fixation, wash 3× with 1× PBS/0.2% triton (PBT) followed by gradual dehydration into 100% ethanol (0/25%/50%/75%/100% steps 5–15 min each depending on the size of the tissue), and keep at −80 °C for at least 2 days before starting the next step, since this is part of the permeabilization process. If needed, samples can be stored at −80 °C for months if not years without significant RNA degradation.
2. Rehydrate the samples by gradually bringing them back to 1× PBS from 100% ethanol (100%/75%/50%/25%/0% with 5–15 min/step), and finally wash 2× with 1× PBT. Bring the samples gradually into 20% sucrose in 1× PBT on a nutator at 4 °C for 3–4 h (*see* **Note 4**).
3. Remove all sucrose carefully and replace it with the O.C.T. embedding compound. Rotate the tubes a few times by hand and let the samples equilibrate at room temperature for 10 min before transferring the samples into cryomolds under a dissection microscope by using a sterile plastic transfer pipette (cut the tip if sample size is big). The O.C.T. compound is very viscous; operate with a relatively large amount (1 ml per sample) to enable the transfer and to avoid bubbles. Once optimally positioned, stiffen the samples first in a flat position by using dry ice before snap freezing them in liquid nitrogen until they start to turn white (submerge the samples ~3–5× for 3 s each). Place the molds at −20 °C for a few minutes (to soften them to enable transfer by using forceps), and collect the samples to individually labeled microcentrifuge tubes for storage at −80 °C.
4. Section the samples into 12–20 μm thick slices by using a cryomicrotome and collect them on AS-coated cover glasses (*see* Subheading 2.2). Depending on the sample size, you may want to try to collect the samples in pairs that will fit into the same hybridization chamber. Store the samples at 4 °C; they can be used for further experiments the next day or up to a few weeks.

3.2 Hybridization and Immunostaining

3.2.1 Hybridization

All the following steps are performed to samples encapsulated inside hybridization chambers. Once the experiment has been started, the samples should be protected from drying. The volume pipetted into each chamber can vary depending on the chamber size (~50 μl for the 8 mm chambers). Several samples can be operated simultaneously, and with the possibility of sample damage or detachment from the cover slip during the long protocol, it may be wise to start the experiment with at least double the amount of samples than what you wish to use for the analysis.

1. Wash the samples three times with 1% PFA 1× PBS solution by slowly rotating them for 5 min. This step removes the O.C.T. compound matrix covering the sample and mounts them tightly onto a cover glass.
2. Permeabilize the samples with 0.5% sodium dodecyl sulfate (SDS) in 1× PBS solution by slowly rotating them for 5 min.
3. Post-fix the samples with 1% PFA in 1× PBS for 5 min.
4. Treat the samples with 1% $NaBH_4$ in 1× PBS solution for 5 min to minimize the background autofluorescence (*see* **Note 2**), and then carefully wash three times with 2× SSC by slowly rotating for 5 min.
5. Block the samples with 1 μM random 60-mer oligonucleotide in 2× SSC for 1 h.
6. Make the hybridization buffer: 10% DS/30% FA/2× SSC with 0.02% thiomersal with 1 μM random 60-mer oligonucleotide. Prepare the probe sets of the current hybridization round by making a mix in the hybridization buffer that contains each probe in a final concentration of 1 nM. Depending on how many channels are used for the imaging, up to five different initiators can be used simultaneously in the mix.
   (a) Gene A–B1
   (b) Gene B–B2
   (c) Gene C–B3
   (d) Gene D–B4
   (e) Gene E–B5
7. Pipette the hybridization solution into the hybridization chambers and place the cover glasses into a humidified chamber (*see* **Note 5**) that is covered with laboratory parafilm. Incubate the samples in 100% humidity at 37 °C for overnight (minimum of 8 h).

3.2.2 Amplification and Probe Stripping (See Fig. 1)

1. Wash the samples three times with 30% FA 2× SSC followed by three times with 2× SSC after the overnight hybridization (*see* **Note 6**).
2. Snap heat the hairpins that will be used for signal amplification (B1H1, B1H2, B2H1, B2H2, etc.) for 90 s in separate microcentrifuge tubes at 95 °C by using a heat block followed by a cool down at room temperature for 30 min. The "standard dye–hairpin set" we have used in our experiments are the following: Cy7-B1, Alexa647-B2, Alexa594-B3, Cy3B-B4, and Alexa488-B5.

3. Mix the hairpins to a final concentration of 120 nM per hairpin in 10% DS/2× SSC solution, pipette it to the hybridization chambers, and incubate by slowly rotating them for 1.5 h in the dark (*see* **Note 7**). After this step, the samples should be protected from light until they are imaged.
4. Wash three times with 30% FA/2× SSC and three times with 2× SSC (*see* **Note 6**). The samples are now ready for imaging. If nuclear staining is desired, a 10-min 1× DAPI/2× SSC staining can be done between the three sets of washes.
5. Since individual transcripts are imaged as diffraction-limited dots, the use of anti-bleach buffer is highly recommended especially when photo-unstable fluorescent dyes are used. Mix the components of the anti-bleach buffer immediately before imaging (the solution is effective only for a couple of hours) with the following ratios: 7 volumes of ABBB, 1 volume of 8% D-glucose, 1 volume of pyranose oxidase, and 1 volume catalase enzyme. Pipette the mixed solution into hybridization chambers, cover them with another cover glass to prevent evaporation, and start imaging.
6. After imaging, wash the samples twice with 2× SSC. Prepare the DNase I solution according the manufacturer's specifications with the protein concentration of 500 units/ml. Add DNase to the chambers and incubate for 1 h in order to strip the probes.
7. Wash the samples three times with 30% FA/2× SSC and three times with 2× SSC (*see* **Note 8**).
8. Now, the samples are ready for the next hybridization round, and the protocol can be repeated from **step 5** in Subheading 3.2.1 (1 μM random 60-mer oligonucleotide blocking in 2× SSC for 1 h) with a new set of genes (*see* **Note 9**). RNA integrity is maintained for weeks, and if necessary, one can take brakes in between the long measurement routine. In that case, the samples should be stored at 4 °C with 2× SSC/0.02% thiomersal and by making sure they do not dry.

3.2.3 Antibody Staining of Plasma Membranes

Achieving single-cell resolution is critical for SGA. For a reliable cell segmentation, immunostaining of proteins localized on the plasma membrane should be used. In order to achieve a strong and uniform signal, it may be necessary to use two or even more antibodies targeting different membrane proteins that are then visualized by using secondary antibodies conjugated with the same fluorophore (e.g., β-catenin together with E-cadherin were used for the dorsal neural tube). Most importantly, for successful data analysis, the target proteins should be chosen based on their expression in the entire tissue of interest to define all cell boundaries. The antibodies should be chosen and tested in the respective tissue after treatment with the SGA sample fixation and hybridization conditions beforehand prior to starting the hybridizations with the actual samples.

1. After the last hybridization and DNA stripping round, wash samples three times with 30% FA/2× SSC and three times with 2× SSC before changing the buffer to 1× PBS. Block the samples for 1 h with the immunostaining blocking solution (5% BSA, and 1% DMSO in 1× PBT).
2. Incubate the samples with primary antibodies in blocking solution overnight at 4 °C while rotating slowly. This step should be performed in a humidified box and the hybridization chambers should be covered with parafilm to prevent evaporation (*see* **Note 9**).
3. The next day, wash the samples 5× 30 min with 1× PBS and then incubate with the secondary antibodies in blocking solution for 3 h at room temperature followed by a wash 5× 30 min with 1× PBS. Finally, apply the anti-bleach buffer before imaging.

3.3 Imaging and Data Analysis

A spinning disc confocal microscope is the most useful microscope for SGA, as it provides the advantage of high-speed and optical sectioning when multiple samples are imaged on multiple channels daily, which can become time consuming. In addition, the resolution and low background of a spinning disc confocal microscope with relatively small photobleaching of fluorophores enables detection of diffraction-limited dots (*see* **Note 10**). It is important to use high-quality oil or water objectives with a magnification that produces ~120–180 nm pixel resolution combined with a high-quantum efficiency CCD or sCMOS camera. Six orthogonal channels (Cy7, Alexa647, Alexa594, Cy3B, Alexa488, and DAPI) that are well separated from each other with high-quality emission and dichroic filter sets are routinely used. The imaging routine is the following:

3.3.1 Imaging

1. Begin by checking the RNA integrity of your sample as follows: choose a highly expressed gene in your tissue of interest for which you have been able to design a set of 24 probes. Design every other probe with one initiator/fluorophore, and the other half with another initiator/fluorophore combination (e.g., odds with Cy7-B1 and evens with Alexa647-B2). Perform the hybridization protocol, image with both channels, and check to what extent the signal from two images with different channels overlaps by counting the percentage of co-localization of the diffraction-limited dots. If the co-localization is >85%, the RNA integrity of the transcripts is acceptable and the experiment can be continued.

2. The N number of hybridization sets with S orthogonal channels are imaged and the DNA probes are stripped between the hybridizations. The number of genes that is measured during an experiment will scale linearly to $N \times S$.
3. To check the RNA integrity after the final hybridization rounds, repeat the probe set used in the first hybridization set and check that the amount of dots in each cell correlate with the images from the first set. You should expect to see a >80% recovery of the signal, and no less than a >60% level is acceptable.
4. As the last imaging step after completion of all the hybridization routines, image the plasma membrane antibody staining that will be used for cell segmentation purposes during the data analysis.

3.3.2 Data Analysis

Data analysis is performed by using Ilastik toolkit [7] (ilastik.org) and MATLAB scripts developed specifically for SGA which can be found at www.singlecellanalysis.org (this website also has a discussion forum and contact information in case help is needed during any step of the SGA protocol or analysis).

1. The Ilastik software is trained to detect plasma membranes and cell interiors, and this information is used to export a probability density map in a .h5 format to present these values in a 2D matrix.
2. Cell interiors from the .h5 file are used in MATLAB as a seed for a watershed algorithm to detect cell boundaries and to convert each cell into 3D volumetric objects in space.
3. Ilastik is trained to recognize the diffraction-limited dots from the images and this data is then exported in .h5 format.
4. 3D cells are aligned with dot images by using a semiautomated alignment routine in the MATLAB program that is based on recognition of easily detectable unique morphological features of each sample.
5. The MATLAB program counts dots (e.g., the number of transcripts of each gene in each cell volume), and stores that data as cell volume-corrected values for further analysis. This way, each cell ends up having volume-corrected values of transcripts for each gene, and that data is stored in a matrix format.
6. The cells are hierarchically clustered based on their gene-expression profile and presented in a heatmap format. The clusters are mapped back to spatial context by visualizing each cell in a defined cluster with the same color (Fig. 2).

4 Notes

1. Trolox makes a saturated solution where some of the solid particles are still visible. This solution should be filtered through a 200-nm pore size filter by using a 60-cc syringe and then aliquoted to microcentrifuge tubes. Adjust ABBB to pH = 8 with HCl or NaOH if needed before filtering.
2. Mixing $NaBH_4$ with 1× PBS produces a lot of bubbles (H2 gas) and the solution should be used immediately after mixing. The solution should not be stored in a container with a closed lid due to a possible explosion hazard that can lead to eye or skin damage.
3. It is important to use a highest possible acetone purity (at least HPLC quality) to prevent solid residue formation on the cover glasses.
4. To create a sucrose gradient, weigh 200 mg of sucrose into a microcentrifuge tube. Have your sample in ~500 μl PBT in another microcentrifuge tube and pour the sucrose crystals into the tube with the embryos. Fill the tube with PBT to final volume of 1 ml. Place the tube on a nutator in a cold room (4–8 °C) and let the crystals gradually dissolve during 3–4 h. It is important to use a nutator that provides vertical rotation, and the crystals will not dissolve if using a flat rotator.
5. An empty pipette tip box with a lid and ~1 cm of water on the bottom can be used as a humidified chamber. The cover glasses are placed on top of the pipette tip holder plate. Make sure that the laboratory parafilm is tightly closing the hybridization chambers, and that the lid is tightly closed.
6. Make sure no leftover hybridization solution is left inside or on top of the hybridization chambers during the washes, which will increase background.
7. A Petri dish with a lid covered with aluminum foil can be used.
8. It is critical in this step to make sure that all the DNase I solution inside and on the hybridization chambers gets properly washed off. Even small residues of the enzyme may have nuclease activity towards the next set of DNA probes during hybridization.
9. We have successfully repeated more than ten hybridization rounds with no sign of significant degradation of RNA.
10. Alternatively, an epifluorescence microscope can be used for imaging, and the signal-to-noise ratio can be improved by using image processing [3].

Acknowledgments

This work was in part funded by the Division of Intramural Research of the National Institute of Dental and Craniofacial Research at the National Institutes of Health, Department of Health and Human Services, as well as grants from the Academy of Finland, Sigrid Juselius Foundation, Ella and George Ehrnrooth's Foundation, Children's Cancer Foundation Väre, and K. Albin Johansson Foundation to LK.

References

1. Tang F, Barbacioru C, Nordman E, Li B, Xu N, Bashkirov VI, Lao K, Surani MA (2010) RNA-Seq analysis to capture the transcriptome landscape of a single cell. Nat Protoc 5 (3):516–535. https://doi.org/10.1038/nprot.2009.236
2. Raj A, van den Bogaard P, Rifkin SA, van Oudenaarden A, Tyagi S (2008) Imaging individual mRNA molecules using multiple singly labeled probes. Nat Methods 5(10):877–879. https://doi.org/10.1038/nmeth.1253
3. Lignell A, Kerosuo L, Streichan SJ, Cai L, Bronner ME (2017) Identification of a neural crest stem cell niche by spatial genomic analysis. Nat Commun 8(1):1830. https://doi.org/10.1038/s41467-017-01561-w
4. Shah S, Lubeck E, Schwarzkopf M, He TF, Greenbaum A, Sohn CH, Lignell A, Choi HM, Gradinaru V, Pierce NA, Cai L (2016) Single-molecule RNA detection at depth by hybridization chain reaction and tissue hydrogel embedding and clearing. Development 143 (15):2862–2867. https://doi.org/10.1242/dev.138560
5. Choi HM, Beck VA, Pierce NA (2014) Next-generation in situ hybridization chain reaction: higher gain, lower cost, greater durability. ACS Nano 8(5):4284–4294. https://doi.org/10.1021/nn405717p
6. Lubeck E, Coskun AF, Zhiyentayev T, Ahmad M, Cai L (2014) Single-cell in situ RNA profiling by sequential hybridization. Nat Methods 11(4):360–361. https://doi.org/10.1038/nmeth.2892
7. Sommer C, Straehle C, Kothe U, Hamprecht FA (2011) Ilastik: interactive learning and segmentation toolkit. IEEE international symposium on biomedical imaging. pp 230–233

Methods in Molecular Biology (2019) 2002: 165–180
DOI 10.1007/7651_2018_185

Published online: 23 September 2018

3D Culture of Mesenchymal Stem Cells in Alginate Hydrogels

Sílvia J. Bidarra and Cristina C. Barrias

Abstract

Three-dimensional (3D) cell culture systems have gained increasing interest among the scientific community, as they are more biologically relevant than traditional two-dimensional (2D) monolayer cultures. Alginate hydrogels can be formed under cytocompatibility conditions, being among the most widely used cell-entrapment 3D matrices. They recapitulate key structural features of the natural extracellular matrix and can be bio-functionalized with bioactive moieties, such as peptides, to specifically modulate cell behavior. Moreover, alginate viscoelastic properties can be tuned to match those of different types of native tissues. Ionic alginate hydrogels are transparent, allowing routine monitoring of entrapped cells, and crosslinking can be reverted using chelating agents for easy cell recovery. In this chapter, we describe some key steps to establish and characterize 3D cultures of mesenchymal stem cells using alginate hydrogels.

Keywords 3D cell culture, Alginate, Artificial matrix, Bioengineered 3D matrix, Hydrogel, Mesenchymal stem cells (MSC), Peptide-modified alginate

1 Introduction

In native tissues, the extracellular matrix (ECM) is a major component of the cellular microenvironment, not only providing structural and mechanical support but also playing an active role in regulating cell behavior and function [1]. The ECM acts as a reservoir of signaling molecules, such as growth factors, and conveys biochemical/biophysical cues via specific cell receptors [1]. It is highly dynamic, being amenable to cell-driven synthesis, degradation, and reassembly over time.

The partial recreation of in vivo environments in a lab setting is fundamental for studying cellular behavior and response to different stimuli, under more defined conditions [2]. Therefore, in vitro cell culture models stand out as essential tools, not only for fundamental cell biology studies but also for testing new therapeutic strategies [3, 4]. While classical methodologies involve the use of two-dimensional (2D) monolayer cultures, it is currently accepted that 2D models are too simplistic [5]. They are built on materials with biophysical properties very different from those of native tissues, fail to provide a three-dimensional (3D) architecture to cells, and lack the means to recapitulate the spatial/temporal changes of the ECM [2]. Thus, although these approaches are widespread and have significantly improved our understanding of cellular behavior, there is increasing evidence that 2D systems can result in cellular activities that deviate significantly from those observed in vivo [6]. Consequently, the paradigm shift from 2D to 3D culture is underway and progressing rapidly. In this context, different biomaterials, particularly hydrogels, are being explored as ECM mimics for cell entrapment and 3D cell culture.

Hydrogels are crosslinked polymeric 3D networks, highly hydrated and compliant, intrinsically recapitulating key features of the natural ECM. They are also permeable, allowing for an efficient exchange of nutrients, oxygen, and cellular metabolites with the extracellular milieu, essential for the survival of entrapped cells [7]. Due to a number of attractive features, alginate hydrogels are among the most frequently used cell-entrapment 3D matrices [8–10]. Alginates are natural polysaccharides that can be extracted from brown algae or produced by bacteria. The polymer consists of linear macromolecules composed by 1,4-linked β-D-mannuronic acid (M) and α-L-guluronic acid (G) monomers, arranged into M-blocks, G-blocks, and/or MG-blocks [11]. Alginate hydrogels can be formed by both chemical and physical crosslinking, being ionic-crosslinked hydrogels the most widely used, as they can be easily made under mild and cytocompatible conditions. Ionic crosslinking results from electrostatic interactions between divalent cations, such as calcium ions, and negatively charged carboxylic groups present in the polymer. These interactions are primarily established between G-blocks, which present a specific spatial arrangement that favors interchain binding as described by the "egg-box" model (Fig. 1a) [12]. Alginate hydrogels are highly versatile, since different strategies can be used to modulate their biochemical and biophysical properties [13]. The viscoelastic properties of alginate solutions and their hydrogels can be tuned, namely by changing the polymer molecular weight, composition (G/M ratio), and/or concentration. Also, functional groups present in alginate chains offer the potential for covalent modification by different chemical routes. While cells are not able to specifically interact with the alginate network, "bioactive" alginate derivatives can be easily obtained by chemical grafting of cell-instructive moieties, such as peptides. Alginate hydrogels modified with different types of peptides have already been described, namely with integrin-binding cell adhesive peptides [14–22], protease-sensitive peptides [23, 24], and osteoconductive peptides [17]. Modification of otherwise "bioinert" alginate hydrogels with integrin-binding-peptides, such as arginine–glycine–aspartic acid (RGD), is often recurrent, not only to promote cell–matrix adhesion, a key requirement for the survival of anchorage-dependent cells, but also to mediate cell–matrix crosstalk.

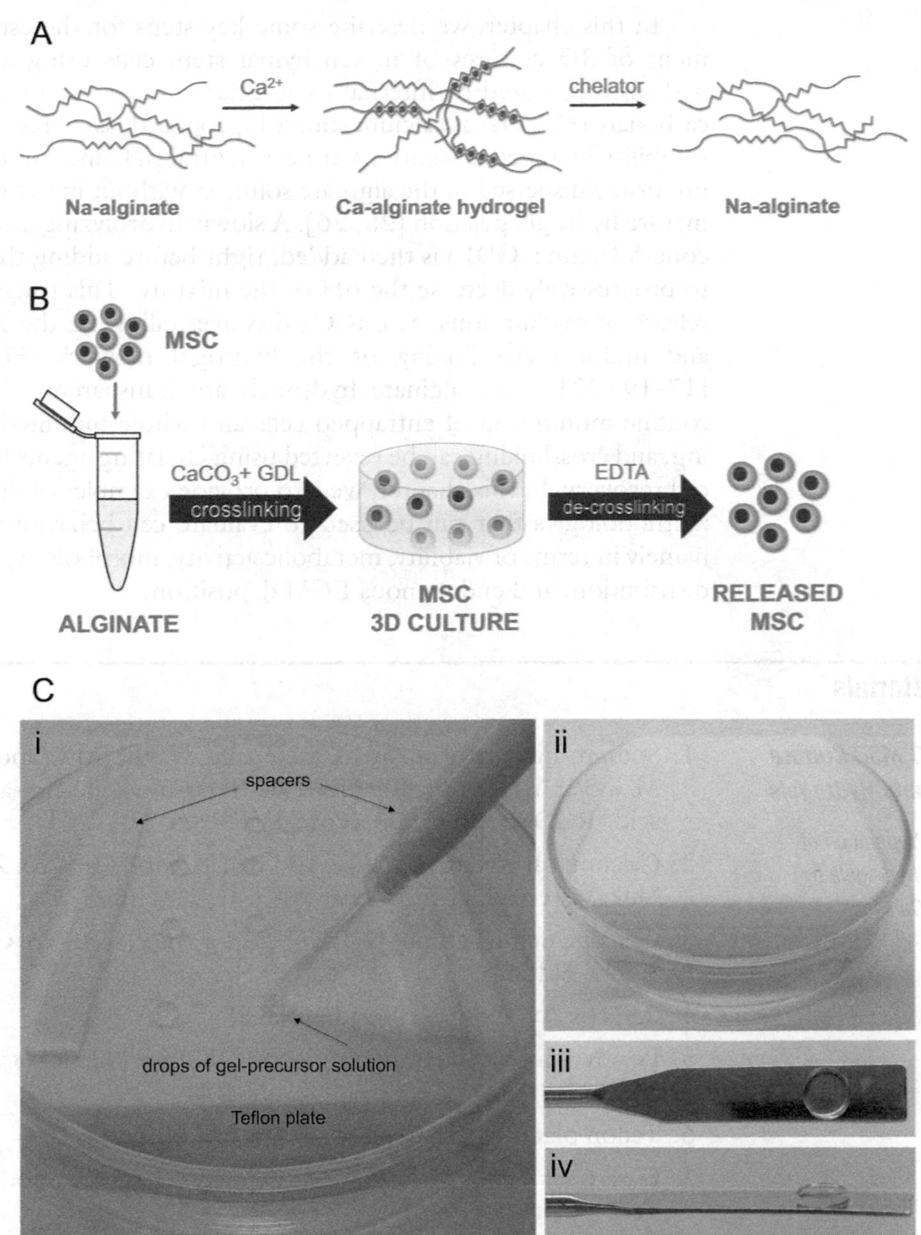

Fig. 1 (**a**) Schematic representation of alginate crosslinking in the presence of divalent cations, such as calcium, according to the "egg-box" model. Alginate crosslinking can be reverted using chelators. (**b**) MSC embedded in calcium alginate hydrogels prepared by internal ionic gelation. Alginate hydrogels can be formed in situ upon release of calcium ions from an insoluble compound (calcium carbonate), triggered by the pH decrease resulting from glucone-δ-lactone (GDL) hydrolysis. Cells can be recovered from the hydrogel using a calcium chelator like ethylenediaminetetraacetic acid (EDTA). (**c**) (i) Alginate discs can be obtained by dispensing drops of gel-precursor solution on a Teflon plate, using a positive-displacement micropipette. (ii) Gels are casted between two Teflon plates with spacers of adequate thickness. (iii and iv) After crosslinking, disc-shaped transparent hydrogels are formed

In this chapter, we describe some key steps for the establishment of 3D cultures of mesenchymal stem cells using alginate hydrogels prepared by internal ionic gelation. The use of calcium carbonate ($CaCO_3$) as calcium source is proposed, as it presents low solubility in aqueous solutions at near neutral pH, and can thus be uniformly dispersed in the alginate solution without initiating premature hydrogel gelation [25, 26]. A slowly hydrolyzing acid (glucone-δ-lactone, GDL) is then added, right before adding the cells, to progressively decrease the pH of the mixture. This triggers the release of calcium ions, as $CaCO_3$ dissolves, allowing the gradual and uniform crosslinking of the hydrogel network (Fig. 1b) [17–19, 22]. Ionic alginate hydrogels are transparent, allowing routine monitoring of entrapped cells and whole-mounted imaging, and crosslinking can be reverted using chelating agents for easy cell recovery. In this chapter, we also provide examples of different methodologies that can be used to evaluate cell behavior in 3D, namely in terms of viability, metabolic activity, morphology/spatial distribution, and endogenous ECM deposition.

2 Materials

2.1 3D MSC Culture in Alginate Hydrogels

2.1.1 Preparation of Alginate Hydrogels by Internal Gelation

1. Sodium alginate of adequate molecular weight (Mw) and G-to-M molar ratio, functionalized with arginine–glycine–aspartic acid (RGD) peptides (*see* **Note 1**)
2. Calcium carbonate ($CaCO_3$, 100.071 g/mol, Cat. No. 21060, Fluka), sterilized by dry heat (160 °C, 120 min)
3. Glucone delta-lactone (GDL, 17,814 g/mol, Cat. No. G4750, Sigma-Aldrich)
4. 0.9% (w/v) Sodium chloride (NaCl)
5. Positive-displacement micropipettes and tips (sterile) (*see* **Note 2**)
6. Teflon plates with adequate spacers (*see* **Note 3**)
7. Glass petri dishes with adequate diameter (to accommodate the Teflon plates)

2.1.2 Cells and Media

1. Human mesenchymal stem cells (hMSC, e.g., Lonza)
2. Dulbecco's modified Eagle medium (DMEM), high glucose, supplemented with GlutaMAX (Thermo Fisher Scientific)
3. Heat-inactivated fetal bovine serum (FBS, Thermo Fisher Scientific)
4. Penicillin–streptomycin 100× (P/S, 10,000 units/mL of penicillin and 10,000 μg/mL of streptomycin, Thermo Fisher Scientific)

5. Basal media formulation: DMEM supplemented with 10% (v/v) FBS and P/S (1×)
6. Osteogenic media formulation: DMEM supplemented with 10% (v/v) FBS, P/S (1×), 100 nM dexamethasone (Cat. No. D8893, Sigma-Aldrich), 10 mM ß-glycerophosphate (Cat. No. G9891, Sigma-Aldrich), and 0.05 mM 2-phosphpo-L-ascorbic acid (Cat. No. 49752, Sigma-Aldrich)
7. 0.05% Trypsin solution with phenol red (Thermo Fisher Scientific)
8. Centrifuge
9. Laminar flow hood
10. Incubator (humidified atmosphere with 5% CO_2, 37 °C)

2.2 Assays for Characterization of Cell Behavior in 3D

2.2.1 Viability Assay

1. Trypsin/EDTA solution (0.25% (w/v) trypsin, 50 mM EDTA, and 0.1% (w/v) glucose)
2. 0.4% (w/v) Trypan blue solution in phosphate-buffered saline solution (PBS)
3. Neubauer chamber
4. Optical microscope

2.2.2 Metabolic Activity Assay

1. Resazurin stock solution (0.1 mg/mL in TBS)
2. Black-walled, clear-bottom 96-well microplates (for fluorescence assays)
3. Microplate fluorescence reader ($\lambda_{Ex} = 530 \pm 9$ nm, $\lambda_{Em} = 590 \pm 9$ nm)

2.2.3 Immunostaining (Whole-Mounted Samples)

1. Tris-buffered saline solution (TBS, 50 mM Tris in 150 mM NaCl, pH 7.4)
2. TBS-Ca (TBS with 7.5 mM calcium chloride)
3. Permeabilizing solution: 0.2% (v/v) Triton X-100 in TBS-Ca
4. Fixing solution: 4% (w/v) paraformaldehyde (PFA) in TBS-Ca
5. Blocking solution: 1% (w/v) bovine serum albumin in TBS-Ca
6. F-Actin staining: Alexa Fluor-Phalloidin 488 (Cat. No. A12379, Thermo Fisher Scientific)
7. Fibronectin staining: rabbit anti-fibronectin primary antibody (Cat. No. F3648, Sigma-Aldrich), anti-rabbit secondary antibody: Alexa Fluor 568 (Cat. No. A11011, Thermo Fisher Scientific)
8. DAPI (4′,6-diamidino-2-phenylindole, Cat. No. D9542, Sigma-Aldrich)
9. Confocal laser scanning microscope

2.2.4 Histology and Histochemistry (Paraffin Sections)

1. Absolute ethanol and sequential dilutions in water (50, 70, 80, 90, 95, 96, and 100%)
2. Paraffin
3. Xylene
4. Gill's hematoxylin
5. Eosin Y
6. 1.5% Safranin solution
7. 0.4% Light green
8. Clear-Rite™
9. Entellan® mounting medium
10. APES-coated slides
11. Distilled and deionized water (dd water)
12. 1% Acetic acid
13. Semiautomated microtome (Leica RM2255)
14. Modular embedding system (Microm STP 120-1)

2.2.5 Analysis of Osteogenic Differentiation by ALP Staining (Whole-Mounted Samples)

1. Fast Violet B aliquots (e.g., 2 mL): prepared by dissolving 12 mg of Fast Violet B (Cat. No. 855, Sigma) in 48 mL of water, stored at −20 °C (protected from light)
2. Naphthol AS-MX phosphatase alkaline solution (0.25% w/v solution, Cat. No. 851, Sigma)
3. Optical microscope and/or stereoscope

3 Methods

3.1 3D hMSC Culture in Alginate Hydrogels

3.1.1 Preparation of Alginate Solution

1. To prepare hydrogel-precursor solution, first select the adequate type of alginate and dissolve it at the desired concentration in 0.9% (w/v) NaCl (*see* **Note 1**).
2. Leave stirring overnight (ON) at 4 °C.
3. Sterilize the final alginate solution by filtration (0.22 μm) (*see* **Note 4**).

3.1.2 Preparation of Cell-Laden Alginate Hydrogels

1. Harvest hMSC from the culture flask using trypsin solution.
2. Count the cells and centrifuge (218 × *g*, 5 min) an adequate volume of cell suspension into an Eppendorf tube (1.5 mL), to obtain a cell pellet containing the exact amount of cells that you intend to use (e.g., per mL of alginate) (*see* **Note 5**).
3. Carefully discard the supernatant and reserve the cell pellet (*see* **Notes 6** and 7).
4. Add an appropriate amount of $CaCO_3$ to 0.9% (w/v) NaCl and vortex (*see* **Notes 7–9**).

5. Add the suspension of $CaCO_3$ to the alginate solution, at Ca^{2+}/COO^- molar ratio of 0.288, and vortex well (*see* **Notes 7–9**).
6. Add freshly prepared and sterile-filtered (0.22 μm) GDL solution to the alginate/$CaCO_3$ mixture, at Ca^{2+}/GDL molar ratio = 0.125, and vortex well (*see* **Note 10**).
7. Add the alginate/$CaCO_3$/GDL mixture to the cell pellet and mix carefully (*see* **Notes 6** and **7**).
8. To cast small cell-laden discs, load the mixture on a Teflon plate (e.g., several drops of 20 μL), use spacers of adequate thickness (e.g., 500 μm or 750 μm), and carefully place another Teflon plate on top (*see* **Note 11**) (Fig. 1c).
9. Allow the mixture to crosslink in a CO_2 incubator (*see* **Note 12**).
10. After crosslinking, carefully remove the top plate and add one drop of medium to each disc to facilitate detachment.
11. Transfer the discs with a spatula into a 24-well cell culture plate (preferably a non-treated culture plate, for suspension cells, to prevent released cells from attaching to the bottom of the plate), and add 500 μL of fresh medium. Place the plates in a CO_2 incubator.
12. To induce osteogenic differentiation, add 500 μL of osteogenic medium to each well at a pre-defined time point.

3.2 Assays for Characterization of Cell Behavior in 3D

3.2.1 Viability (Trypan Blue Dye Exclusion Assay)

The Trypan blue exclusion dye is able to penetrate compromised membranes of non-viable cells, but not of intact live cells, staining them in blue. After counting the number of stained vs. total cells, the percentage of viable cells can be calculated.

1. Transfer the discs to a 1.5-mL Eppendorf tube.
2. To dissolve the disc and disaggregate the cells, add 50 μL of sterile trypsin/EDTA (5 min). Do up-and-down with a positive-displacement pipette to mix well, without incorporating air bubbles (*see* **Note 13**).
3. Transfer 10 μL of cell suspension into a 1.5-mL Eppendorf tube and add 10 μL of Trypan blue.
4. Transfer to Neubauer chamber and count stained and total cells. Calculate the percentage of viable cells (*see* **Note 14**).

3.2.2 Metabolic Activity (Resazurin Assay)

Resazurin solution is added to the culture medium and is reduced by metabolically active cells, resulting in the decrease of the oxidized form (resazurin, blue, and nonfluorescent) and concomitantly increase of the reduced form (resorufin, red, and fluorescent). The intensity of resorufin dye can be quantified either by colorimetric (absorbance) or fluorimetric (fluorescence) measurements, but the latter provides higher sensitivity. Generally described as nontoxic to cells, the resazurin assay allows multiple measurements (different time points) on the same sample, although continuous exposure to the dye may ultimately compromise cell viability.

1. Prepare resazurin working solution at 20% (v/v), by diluting the stock solution in complete medium.
2. Transfer the hydrogels to a new 24-well cell culture plate.
3. Add the resazurin working solution (500 μL/well), and incubate plates at 37 °C for 2–3 h, protected from light (*see* **Note 15**).
4. Transfer 200 μL of the supernatants into black-walled, clear-bottom 96-well microplates (for fluorescence assays) and read the fluorescence using a microplate spectrofluorometer ($\lambda_{Ex} = 530 \pm 9$ nm, $\lambda_{Em} = 590 \pm 9$ nm).
5. After the assay, cell-laden hydrogel can be processed for other assays (e.g., viability assay, confocal microscopy), or fresh media can be added and samples re-incubated at 37 °C (*see* **Note 16**).

3.2.3 Immunostaining (Whole-Mounted Samples)

As an example, this protocol describes a procedure for F-actin (cytoskeleton) and fibronectin (ECM component) staining, but different components can be stained using specific antibodies.

1. Wash cell-laden hydrogels with TBS-Ca (*see* **Note 17**).
2. Fix samples with 4% (v/v) PFA in TBS-Ca for 20 min.
3. Wash three times with TBS-Ca.
4. Permeabilize with 0.2% (v/v) Triton X-100 in TBS-Ca for 10 min.
5. Wash three times with TBS-Ca.
6. Incubate with blocking buffer for 1 h at room temperature (RT).
7. Incubate with rabbit anti-human fibronectin primary antibody (1:200, in blocking buffer) for fibronectin staining, and with Alexa Fluor-Phalloidin 488 (1:40) for F-actin staining, ON at 4 °C.
8. Wash three times with TBS-Ca.
9. Incubate with the secondary antibody Alexa Fluor 568 (1:1000, in blocking buffer) for 1 h at RT.
10. Wash three times with TBS-Ca.
11. For nuclei staining incubate with DAPI (0.5 μg/mL in TBS-Ca) for 15 min at RT.
12. Wash three times with TBS-Ca.
13. Capture images using a confocal laser microscope (Fig. 2a) (*see* **Note 18**).

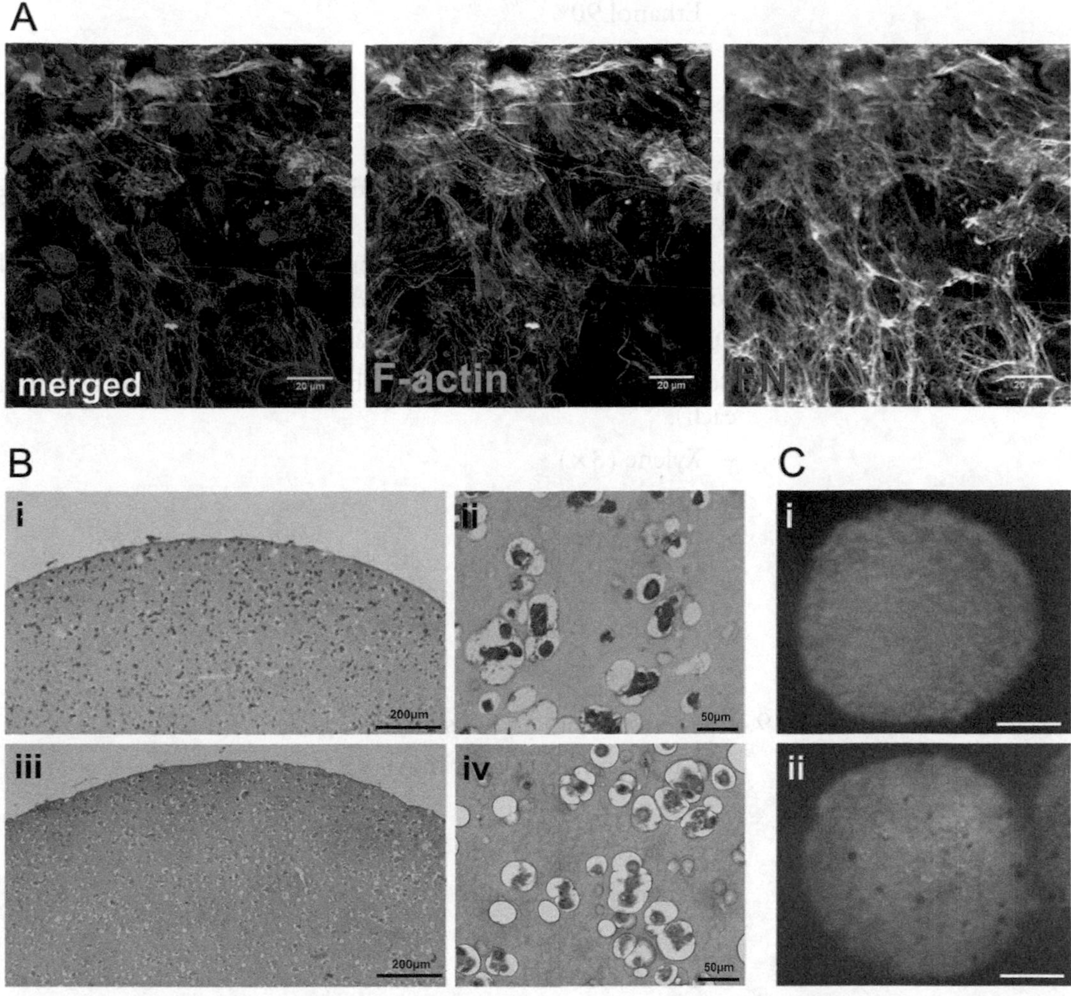

Fig. 2 (**a**) Cell morphology and FN matrix assembly in MSC-laden RGD-alginate hydrogels (1 wt.% alginate, 200 μM RGD). Whole-mounted samples (24 h) were fixed, immunostained, and analyzed by CLSM, as described in the protocol (FN in red, F-actin in green, and nuclei in blue; scale bar: 20 μm). (**b**) (i, ii) Hematoxylin and eosin staining and (iii, iv) safranin O/light green staining of paraffin-embedded MSC-laden hydrogels showing cell spatial distribution (i, ii: alginate stains light pink; iii, iv: alginate stains orange. Scale bars: (i, iii) 200 μm; (ii, iv) 50 μm). (**c**) Alkaline phosphatase (ALP)-stained whole-mounted samples of MSC-laden hydrogels under (i) basal conditions and (ii) osteoinductive conditions, after 2 weeks of culture. MSC express high levels of ALP activity (pink staining) when cultured under osteoinductive conditions (scale bar: 500 μm)

3.2.4 Histology and Histochemistry (Paraffin Sections)

1. Wash the cell-laden hydrogels with TBS-Ca.
2. Fix samples with 4% (v/v) PFA in TBS-Ca for 20 min.
3. Wash with TBS-Ca.
4. Samples are processed in the automated tissue processor with the following steps (20 min each):
   - Ethanol 70%
   - Ethanol 80%

- Ethanol 90%
- Ethanol 95%
- Ethanol100% (2×)
- Clear-Rite (3×)

5. Discs are embedded in paraffin in a modular embedding system.
6. Cut 3 μm sections using a semiautomated microtome and collect the sections on labelled APES-coated slides.
7. Dry the slides at 37 °C during 24 h.
8. Before staining, sections must be dewaxed as follows (3 min each):
   - Xylene (3×)
   - Ethanol 100% (2×)
   - Ethanol 95%
   - Ethanol 70%
   - Ethanol 50%
   - dd water
9. For hematoxylin and eosin staining (*see* **Note 19**):
   - Gill's hematoxylin (3 min)
   - Tap water (6 min)
   - Ethanol 50% (1 min)
   - Ethanol 70% (1 min)
   - Ethanol 95% (1 min)
   - Ethanol 100% (3×, 1 min each)
   - Eosin Y (2 min)
   - Ethanol 100% (3×, 1 min each)
   - Xylene (3×, 3 min each)
10. For safranin O/light green (*see* **Note 20**):
    - Gill's hematoxylin (2 min)
    - Tap water (4 min)
    - Light green (5 min)
    - Acetic acid 1% (30 s)
    - Safranin solution (30 min)
    - Ethanol 96% (briefly)
    - Ethanol 100% (briefly)
    - Xylene (3×, 2 min each)
11. Mount the sections with Entellan® mounting medium.
12. Capture images by optical microscopy (Fig. 2b).

3.2.5 Analysis of Osteogenic Differentiation by ALP Staining (See ***Note 21****) (Whole-Mounted Samples)*

1. Wash the cell-laden hydrogels with TBS-Ca.
2. Fix samples with 4% PFA in TBS-Ca for 20 min.
3. Wash 1× with TBS-Ca.
4. Thaw one aliquot of Fast Violet B (2 mL), add 80 μL of 0.25% w/v Naphthol AS-MX, and mix well (if not used immediately, wrap the tube in aluminum foil to protect from light) (*see* **Note 22**).
5. Incubate samples with Naphthol AS-MX phosphate/Fast Violet B solution for 30 min at 37 °C protected from light.
6. Wash 1× with TBS-Ca.
7. Capture images in an optical microscope and/or stereoscope (Fig. 2c).

4 Notes

1. Alginate properties, namely viscoelastic properties and hydrogel-formation ability, are dependent on its Mw and monomer composition (G-to-M ratio and sequence). There are several commercially available alginates, from different sources, with different Mw and G/M ratios. For example, NovaMatrix® sells ultrapure (endotoxin levels lower than 100 EU/g) and well-characterized sodium alginates with different Mw and G/M ratio (PRONOVA™ UP alginates, Drug Master Files are maintained with the US FDA), as well as peptide-coupled alginates (NOVATACH™), allowing customers to select products with consistent properties between different batches. Peptide-modified alginates can also be prepared *in house*, using different types of chemistries (e.g., carbodiimide) as described in [15–17, 21, 23, 27], using custom-made peptides (available from different companies).
2. Alginate solutions are viscous and should always be handled with positive-displacement micropipettes.
3. Teflon plates and spacers can be custom-made. For example, using square plates of 10 × 10 cm^2 and 1 cm thickness allows two plates to perfectly fit in a glass petri dish with a diameter of 150 mm, for easy handling and incubation under sterile conditions. Teflon spacers can be rectangles with 8 cm length, 0.5 cm width, and variable thickness (depending on the desired disc height). In the absence of Teflon plates, hydrogels can be formed by directly pipetting the solution into non-treated (hydrophobic) culture plates, forming drops. After crosslinking, this will yield hydrogels with hemisphere shape rather than cylinders.

4. Low-viscosity alginate solution (with low alginate concentration and/or low Mw) can be sterilized by filtration using a 0.22-μm syringe filter. To prepare high-viscosity solutions, first prepare a low-viscosity solution (generally up to 1% w/v, but this depends on the Mw), then filter using a Steriflip® filter unit and lyophilize. Pre-sterilized alginate should then be dissolved at the desired concentration in sterile NaCl and stirred ON at 4 °C.
5. The behavior of hMSC in 3D alginate cultures is highly dependent on the initial cell density. As demonstrated by Maia et al., higher cell densities not only promoted higher cell–cell contact and secretion of ECM proteins (e.g., fibronectin and collagen) but also stimulated hMSC osteogenic differentiation [22]. So, the final cell density should always be optimized, according to the specific cell type and intended application.
6. Cell pellet should be resuspended in a small volume of 0.9 w/v % NaCl to facilitate mixing with the alginate solution. Importantly, since this volume will affect the final volume of the hydrogel, it must be considered when calculating the final alginate concentration.
7. The volumes of the suspensions of crosslinking agents and cells that are sequentially added to the alginate solution must be taken into account, as they will affect the final alginate concentration. For example, to prepare 1 mL of pre-gel mixture with a final alginate concentration of 1% (w/v), an alginate solution at higher concentration (e.g., 1 mL at 1.7% w/v) should be initially prepared. After addition of the crosslinking agents (1.73 mg of $CaCO_3$ in 50 μL and 24.68 mg of GDL in 258 μL), the alginate concentration decreases to 1.3% (w/v). This intermediate concentration is finally mixed with the cell suspension (1 mL of 1.3% alginate with $CaCO_3$ and GDL is added to 300 μL of cell suspension), yielding a final concentration of 1% (w/v) alginate [17, 27, 28]. Volumes/concentrations should be adjusted according to the application.
8. Ca^{2+} ions participate in the ionic interchain binding between carboxyl groups (COO^-) (mainly from guluronic acid blocks) present in adjacent alginate molecules. The Ca^{2+}/COO^- molar ratio of 0.288 has been optimized, based on the previous studies [16, 25, 27].
9. 1 g of alginate contains 5.053×10^{-3} mol of carboxyl groups. In an alginate with 70% of guluronic acid residues, the total moles of COO^- in those residues is 3.54×10^{-3} mol.
10. The release of Ca^{2+} into the solution is promoted by the generation of an acidic pH with glucone delta-lactone (GDL), a slowly dissociating acid, which is incorporated in the solution. GDL solution should be prepared and filtered

immediately before use since it progressively hydrolyzes forming gluconic acid. If it is already hydrolyzed before addition of alginate, it will immediately trigger crosslinking and nonuniform hydrogels with clumps will be formed [16, 17]. The $CaCO_3$ to GDL molar ratio of 0.125 promotes release of calcium ions from $CaCO_3$ at adequate rate, without compromising cellular viability [16–18, 20, 22].

11. The dimensions of hydrogel discs will depend on the drop volume and the thickness of the spacers, which should be adjusted as desired. For example, for confocal microscopy imaging (to be able to image all the cross section of the disc), a maximum height of ~250 μm is preferable.
12. Hydrogel crosslinking starts immediately after the addition of GDL to the $CaCO_3$/alginate mixture. The gelling time depends on different parameters, such as the type and concentration of alginate and crosslinking agents, and the total amount of cells. It should be optimized for a particular application. For the example described in **Note 7**, the gelling time is typically 20–30 min.
13. The crosslinking of calcium alginate hydrogels can be reverted by adding chelators such as ethylenediaminetetraacetic acid (EDTA) or sodium citrate, which will dissolve the hydrogel and release the entrapped cells. In alternative, the enzyme alginate lyase can be used to cleave the polymeric chains [29]. Trypsin is often added in combination with these agents to disrupt cell–cell and cell–matrix interactions and obtain uniform cell suspensions.
14. Viable cells are seen as bright cells and nonviable cells are stained dark blue.
15. Under most experimental conditions, the fluorescent signal will be proportional to the number of viable cells. However, the final resazurin concentration and the incubation time might need to be optimized according to the characteristics of the cells and sample (e.g., 2D vs. 3D, as in the later diffusional limitations may interfere with the assay). The response of a particular cell type/system (linear range, and lower detection limit) can be tested by measuring the fluorescent signal from discs with different amounts of entrapped cells.
16. Resazurin is considered to be nontoxic to cells. So, if you intend to perform a continuous assay, remove the medium containing resazurin, add fresh medium, and then put the plates back into the CO_2 incubator. To guarantee that all the resazurin is removed from the hydrogel network, renew the medium after 30 min. At any time point, 3D cell-laden hydrogels can be reused for other assays.

17. Stabilization of ionically crosslinked alginate hydrogels in TBS can be achieved by adding free calcium ions to the solution, at an Na:Ca molar ratio of less than 25:1 for high-guluronate alginates and 3:1 for low-guluronate alginates [30]. This will prevent samples from disintegration during handling in the different assays. Important note: PBS should not be used in the assays, as phosphate ions will capture calcium ions leading to sample disintegration.
18. Transfer the discs to a glass bottom dish for higher-resolution imaging.
19. Hematoxylin is a basic/positive dark blue/violet stain that binds to basophilic negatively charged substance such as nucleic acids, while eosin is an acidic/negative red/pink that binds to acidophilic substances such as positively charged amino acids present in proteins [31].
20. Safranin-O is commonly used for staining negatively charged proteoglycans (e.g., from cartilage and mucins) in orange to red, and it can also be used to stain polyanionic macromolecules such as alginate. Sections are counterstained with fast green and hematoxylin to visualize cells and nuclei in bluish green [32].
21. Alkaline phosphatase (ALP) is a general term used to describe nonspecific phosphomonoesterases, which hydrolyze phosphate monoesters at alkaline pH [33]. ALP is found in many tissues but is present at particularly high concentrations in bone, being involved in bone matrix deposition and mineralization. Thus, ALP activity is commonly used as an osteogenic differentiation marker.
22. ALP activity in bone cells can be evaluated by using Naphthol AS-MX Phosphate and Fast Violet B. ALP activity is visualized by the presence of red/pink stained areas.

Acknowledgments

Project 3DEMT funded by POCI-Operacional Programme for Competitiveness and Internationalisation via FEDER-*Fundo Europeu de Desenvolvimento Regional* (POCI-01-0145-FEDER-016627) and by Portuguese Foundation for Science and Technology (FCT) via OE-*Orçamento de Estado* (PTDC/BBB-ECT/251872014). The authors thank FCT the post-doctoral grant SFRH/BPD/80571/2011 (Sílvia J. Bidarra) and the IF research position IF/00296/2015 (Cristina C. Barrias).

References

1. Thomas D, O'Brien T, Pandit A (2018) Toward customized extracellular niche engineering: progress in cell-entrapment technologies. Adv Mater 30(1)
2. Justice BA, Badr NA, Felder RA (2009) 3D cell culture opens new dimensions in cell-based assays. Drug Discov Today 14(1–2):102–107
3. Huang G, Li F, Zhao X et al (2017) Functional and biomimetic materials for engineering of the three-dimensional cell microenvironment. Chem Rev 117(20):12764–12850
4. Edmondson R, Broglie JJ, Adcock AF et al (2014) Three-dimensional cell culture systems and their applications in drug discovery and cell-based biosensors. Assay Drug Dev Technol 12(4):207–218
5. Duval K, Grover H, Han L-H et al (2017) Modeling physiological events in 2D vs. 3D cell culture. Physiology 32(4):266–277
6. Baker BM, Chen CS (2012) Deconstructing the third dimension—how 3D culture microenvironments alter cellular cues. J Cell Sci 125 (Pt 13):3015–3024
7. Tibbitt MW, Anseth KS (2009) Hydrogels as extracellular matrix mimics for 3D cell culture. Biotechnol Bioeng 103(4):655–663
8. Bidarra SJ, Barrias CC, Granja PL (2014) Injectable alginate hydrogels for cell delivery in tissue engineering. Acta Biomater 10 (4):1646–1662
9. Bidarra SJ, Torres AL, Barrias CC (2016) Injectable cell delivery systems based on alginate hydrogels for regenerative therapies. In: Hashmi S (ed) Reference module in materials science and materials engineering. Elsevier, Oxford, pp 1–17. https://doi.org/10.1016/B978-0-12-803581-8.04057-1
10. Lee KY, Mooney DJ (2012) Alginate: properties and biomedical applications. Prog Polym Sci 37(1):106–126
11. Smidsrod O, Skjak-Braek G (1990) Alginate as immobilization matrix for cells. Trends Biotechnol 8(3):71–78
12. Morch YA, Donati I, Strand BL (2006) Effect of Ca2+, Ba2+, and Sr2+ on alginate microbeads. Biomacromolecules 7(5):1471–1480
13. Lee K, Mooney D (2001) Hydrogels for tissue engineering. Chem Rev 101:1869–1879
14. Evangelista MB, Hsiong SX, Fernandes R et al (2007) Upregulation of bone cell differentiation through immobilization within a synthetic extracellular matrix. Biomaterials 28 (25):3644–3655
15. Bidarra SJ, Barrias CC, Barbosa MA et al (2010) Immobilization of human mesenchymal stem cells within RGD-grafted alginate microspheres and assessment of their angiogenic potential. Biomacromolecules 11 (8):1956–1964
16. Bidarra SJ, Barrias CC, Fonseca KB et al (2011) Injectable in situ crosslinkable RGD-modified alginate matrix for endothelial cells delivery. Biomaterials 32(31):7897–7904
17. Maia FR, Barbosa M, Gomes DB et al (2014) Hydrogel depots for local co-delivery of osteoinductive peptides and mesenchymal stem cells. J Control Release 189:158–168
18. Maia FR, Fonseca KB, Rodrigues G et al (2014) Matrix-driven formation of mesenchymal stem cell-extracellular matrix microtissues on soft alginate hydrogels. Acta Biomater 10 (7):3197–3208
19. Torres AL, Bidarra SJ, Pinto MT et al (2018) Guiding morphogenesis in cell-instructive microgels for therapeutic angiogenesis. Biomaterials 154:34–47
20. Bidarra SJ, Oliveira P, Rocha S et al (2016) A 3D in vitro model to explore the interconversion between epithelial and mesenchymal states during EMT and its reversion. Sci Rep 6:27072
21. Rowley JA, Madlambayan G, Mooney DJ (1999) Alginate hydrogels as synthetic extracellular matrix materials. Biomaterials 20 (1):45–53
22. Maia FR, Lourenco AH, Granja PL et al (2014) Effect of cell density on mesenchymal stem cells aggregation in RGD-alginate 3D matrices under osteoinductive conditions. Macromol Biosci 14(6):759–771
23. Fonseca KB, Gomes DB, Lee K et al (2014) Injectable MMP-sensitive alginate hydrogels as hMSC delivery systems. Biomacromolecules 15(1):380–390
24. Fonseca KB, Maia FR, Cuz FA et al (2013) Enzymatic, physiocochemical and biological properties of MMP-sensitive alginate hydrogels. Soft Matter 9:3283–3292
25. Kuo CK, Ma PX (2001) Ionically crosslinked alginate hydrogels as scaffolds for tissue engineering: part 1. Structure, gelation rate and mechanical properties. Biomaterials 22(6):511–521
26. Oliveira SM, Barrias CC, Almeida IF et al (2008) Injectability of a bone filler system based on hydroxyapatite microspheres and a vehicle with in situ gel-forming ability.

J Biomed Mater Res B Appl Biomater 87B (1):49–58

27. Fonseca K, Bidarra SJ, Oliveira MJ et al (2011) Molecularly-designed alginate hydrogels susceptible to local proteolysis as 3D cellular microenvironments. Acta Biomater 7 (4):1674–1682
28. Alsberg E, Kong HJ, Hirano Y et al (2003) Regulating bone formation via controlled scaffold degradation. J Dent Res 82(11):903–908
29. Formo K, Aarstad OA, Skjak-Braek G et al (2014) Lyase-catalyzed degradation of alginate in the gelled state: effect of gelling ions and lyase specificity. Carbohydr Polym 110:100–106
30. D'Ayala G, Malinconico M, Laurienzo P (2008) Marine derived polysaccharides for biomedical applications: chemical modification approaches. Molecules 13(9):2069–2106
31. Fischer AH, Jacobson KA, Rose J et al (2008) Hematoxylin and eosin staining of tissue and cell sections. CSH Protoc 2008:pdb.prot4986
32. Ahmad R, Oprenyeszk F, Sanchez C et al (2015) Chitosan enriched three-dimensional matrix reduces inflammatory and catabolic mediators production by human chondrocytes. PLoS One 10(5):e0128362
33. Sharma U, Pal D, Prasad R (2014) Alkaline phosphatase: an overview. Indian J Clin Biochem 29(3):269–278

Methods in Molecular Biology (2019) 2002: 181–193
DOI 10.1007/7651_2018_196

Published online: 12 December 2018

Isolation and Identification of Murine Bone Marrow-Derived Macrophages and Osteomacs from Neonatal and Adult Mice

Joydeep Ghosh, Safa F. Mohamad, and Edward F. Srour

Abstract

Hematopoietic stem cells (HSCs) are regulated by multiple components of the hematopoietic niche, including bone marrow-derived macrophages and osteomacs. However, both macrophages and osteomacs are phenotypically similar. Thus, specific phenotypic markers are required to differentially identify the effects of osteomacs and bone marrow macrophages on different physiological processes, including hematopoiesis and bone remodeling. Here, we describe a protocol for isolation of murine bone marrow-derived macrophages and osteomacs from neonatal and adult mice and subsequent identification by multiparametric flow cytometry using an 8-color antibody panel.

Keywords Bone digestion, CD166, Hematopoietic niche, Bone-marrow macrophages, Multiparameter flow cytometry, Osteomacs

1 Introduction

Macrophages are myeloid cells and part of the innate immune system [1]. Macrophages also act as regulators of hematopoiesis and are required for maintenance of hematopoiesis [2]. Osteomacs are resident tissue macrophages and are part of the endosteal and periosteal bone lining tissues and have identical phenotypical markers of bone marrow-derived macrophages, including CD169 [3]. Osteomacs increase deposition of bone matrix during wound healing and also interact with components of hematopoietic niche, including osteoblasts and megakaryocytes, to regulate hematopoietic stem and progenitor cell function [3, 4]. Depletion of both bone marrow macrophages and osteomacs results in a loss of osteoblasts in bone and mobilization of hematopoietic stem and progenitor cells (HSPC), thus making them critical cellular components of hematopoietic niche [5]. Activated leukocyte cell adhesion molecule (Alcam, CD166), a 65 kDa protein, is a member of the immunoglobulin superfamily and part of a subgroup with five extracellular immunoglobulin-like domains [6–8]. We have previously identified that CD166 is a marker for hematopoietic stem cells and osteoblasts [9, 10]. Recently, we have demonstrated that based on CD166 expression, osteomacs and bone marrow-derived macrophages can be distinguished both in neonatal and adult mice. However, although CD166 can distinguish between osteomacs and macrophages, it is expressed only on a very small subset of osteomacs.

Flow cytometry measures the physical properties (size, granularity) as well as fluorescent properties of single cells. Cell samples, stained with different fluorochrome-conjugated antibodies or dyes, are ran as a single-cell stream through a set of lasers. The electrons in fluorescent probes are excited when they pass through the lasers and enter a higher energy state followed by a return to ground state. During this process, they emit light energy at a longer wavelength. Thus, using multiple lasers and fluorochromes with different emission spectra, flow cytometry can identify multiple parameters from a single sample. Currently with custom-designed flow cytometers with multiple lasers, it is possible to create a panel of 20 or more fluorochromes [11, 12]. For murine bone marrow-derived macrophages, the single-cell suspension required for flow cytometry is obtained by flushing or crushing long bones [3]. For osteomacs, the cells have to dissociate from bone which requires enzymatic digestion to create a single-cell suspension [3]. Enzymatic digestion of tissues might affect expression of cell surface markers [13]. However, using our method, we are able to detect all the cell surface proteins in osteomacs, which are similar to bone marrow-derived macrophages. Moreover, we also detected CD166 on osteomacs, a marker to distinguish them from bone marrow-derived macrophages.

The purpose of this chapter is to outline the methods to isolate and identify bone marrow-derived macrophages and osteomacs from neonatal and adult mice using flow cytometry. This staining protocol allows us to distinguish between osteomacs and bone marrow-derived macrophages via the expression of CD166 using an 8-color fluorochrome- conjugated antibody panel.

2 Materials

2.1 Mice

In our laboratory, we have used 8–12-week-old C57BL/6J mice for isolation of bone marrow macrophages and osteomacs from adult mice. For isolation of bone marrow macrophages and osteomacs from neonatal mice, we use 2–3-day-old C57BL/6J pups.

2.2 Reagents and Supplies

This protocol describes isolation of cells from multiple tissue types. Due to the diversity of the tissues to be processed, the requirements for reagents and supplies change for type of tissues. Each section of the protocol shares some common reagents and supplies with other section(s) and also has a unique set of reagents. For convenience, we are mentioning all the materials required for each individual section.

2.2.1 For Isolation of Bone Marrow Macrophages Adult Mice

1. Complete Iscove's Modified Dulbecco's Medium (IMDM). In 445 mL of IMDM, add 50 mL of heat-inactivated fetal bovine serum and 5 mL of penicillin-streptomycin, and filter the medium.
2. 70% ethanol (EtOH).
3. Sterile forceps.
4. Sterile surgical scissors.
5. 70 μm cell strainers.
6. Sterile gauze sponge.
7. 1× red blood cell (RBC) lysis buffer.
8. 10 mL syringes.
9. 25G, 5/8 in. needles.

2.2.2 For Isolation of Osteomacs (from Neonates and Adults)

1. Complete MEM-alpha medium. In 445 mL of MEM-alpha, add 50 mL of heat-inactivated fetal bovine serum and 5 mL of penicillin-streptomycin and filter the medium.
2. Collagenase type 2 (Cat# LS004177, Worthington Biochemical).
3. N^{α}-Tosyl-Lys chloromethyl ketone, hydrochloride (TLCK, Cat# 616382 Millipore Sigma).
   (a) Prepare 100 mM stock solution of TLCK.
   (b) Measure collagenase (22 units/mL) according to the required volume. Add required volume of PBS, vortex well, and add TLCK (1:1,000). The final concentration of TLCK will be 100 μM. Filter the collagenase solution using a steriflip with 40 μm pore size filter (*see* **Note 1**).

2.2.3 For Isolation of Osteomacs from Adults

1. **Items 1–6** in Subheading 2.2.1.

2.2.4 For Isolation of Osteomacs from Neonates

1. T-pins.
2. Ethylenediaminetetraacetic acid (EDTA). Prepare 4 mM of EDTA solution prior to the experiment and prewarm in a 37 °C water bath.

2.2.5 For Isolation of Bone Marrow Macrophages from Neonates

1. Sterile porcelain mortar and pestle.
2. **Items 1** to 7 in Subheading 2.2.1.

2.3 For Immunostaining

1. Stain-wash buffer: 1× phosphate-buffered saline (PBS) supplemented with 1% fetal bovine serum and 1% penicillin-streptomycin. Store it at 4 °C.
2. TruStain FcX (anti-mouse CD16/32; clone 93).

3. Pacific Blue anti-mouse CD45 (clone 30-F11), allophycocyanin (APC) F4/80 (clone BM8), APC-Cy7 CD11b (clone M1/70), PerCP-Cy5.5 CD68 (clone FA-11), Alexa Fluor 488 Mac2 (clone M3/38), PE Dazzle 594 Ly-6G (clone 1A8), PE-Cy7 M-CSFR (clone AFS98), and R-phycoerythrin (PE) CD166 (clone eBioALC48).
4. 5 mL round-bottom polystyrene tubes with 40 μm cell strainer caps.

2.4 For Flow Cytometry Acquisition and Analysis

1. BD LSRFortessa or BD LSR II flow cytometer (BD Biosciences; Franklin Lakes, NJ).
2. FlowJo software (FlowJo, LLC; Ashland, OR).

3 Methods

Subheadings 3.1, 3.3, and 3.4 can be followed individually to isolate the respective cells described in the specific section. To isolate osteomacs from adult mice, as described in Subheading 3.2, complete **steps 1–6** of Subheading 3.1.

3.1 Isolation of Bone Marrow Macrophages from Adults

1. The complete IMDM medium and 1× RBC lysis buffer should be brought to room temperature prior to use.
2. On a biosafety cabinet, place an absorbent towel and spray 70% EtOH on it generously.
3. Euthanize 8–12-week-old C57BL/6J mouse according to institution- approved protocol. Spray the mouse with 70% EtOH before taking it inside the biosafety cabinet.
4. Isolate two tibias and two femurs from the hind limbs and keep them in a sterile cell culture plate containing complete IMDM. Keep the epiphyses intact while isolating the bones.
5. Using a sterile gauze sponge, strip the soft tissues and muscles from the bones and keep the bones in a new sterile cell culture plate containing complete IMDM.
6. Collect complete IMDM in a 10 mL syringe with a 25G needle.

   Cut the tibias and femurs in half using the sterile scissors. Grab the end of the bone using sterile forceps and insert the tip of the needle into the shaft of the bone. Slowly inject 1–2 mL of the IMDM while slowly retracting and pushing the needle toward the epiphyses. To ensure that the epiphyses are washed with the medium, introduce the tip slowly near the epiphyses while flushing. Inject medium as long as the bone stays red. If the flushing of bone marrow is complete, the bones would look pale.

7. Using a 70 μm cell strainer, filter the cells into a fresh 50 mL centrifuge tube.
8. Centrifuge the tube at 500 × *g* for 6 min at 4 °C and discard the supernatant.
9. Resuspend the cell pellet in 1 mL of 1× RBC lysis buffer at room temperature for 3–4 min.
10. Add 10 mL of complete media to the tube, centrifuge the 50 mL tube at 500 × *g* for 6 min at 4 °C, and discard the supernatant.
11. If the lysis is incomplete, the cell pellet will appear reddish. Repeat **steps 9** and **10** once.
12. Resuspend the cells in 5 mL of stain-wash buffer.
13. Count the cells using a hemocytometer or automated cell counter.
14. Resuspend the cells in stain-wash buffer at a final concentration of 2.5×10^7/mL.

3.2 Isolation of Osteomacs from Adults

1. Prepare the collagenase solution and keep it in a 37 °C water bath for 30 min before use.
2. The protocol described below is for one mouse. Bones from three to four mice can be pooled in a single tube.
3. In a 50 mL sterile tube, add 15 mL complete MEM-alpha medium. Keep the tube on ice with a 70 μm cell strainer on top of it inside a biosafety cabinet.
4. Isolate and flush the bones as described in Subheading 3.1 (**steps 1–6**).
5. After flushing the bones, cut the bones into small pieces, approximately 1 mm long.
6. In a 15 mL tube, collect the bone pieces in 10 mL of PBS and vortex the tube twice. After vortexing, discard the PBS with a pipette carefully. Do not take out the bone pieces while discarding the PBS.
7. Add 1.5 mL of collagenase to the tube containing bone pieces and keep it in a 37 °C water bath shaker for 15 min. Vortex the tube at an interval of 5 min.
8. After 15 min, using a 1 mL pipette, collect the collagenase supernatant into the 50 mL tube kept on ice.
9. Add 2 mL of complete MEM-alpha medium on the strainer to rinse it.
10. Repeat **steps** 7 to **9** for four more times for a total of five digestions.
11. After five digestions, centrifuge the 50 mL tube at 500 × *g* at 4 °C and discard the supernatant.

12. Resuspend the cells in 8 mL stain-wash buffer and filter through a 70 μm cell strainer into a new 50 mL centrifuge tube.
13. Centrifuge the 50 mL tube at 500 × *g* at 4 °C and discard the supernatant.
14. Resuspend the cells in 300 μL of stain-wash buffer.
15. Count the cells using a hemocytometer or an automated cell counter.
16. Resuspend the cells in stain-wash buffer at a final concentration of 2.5×10^7/mL.
17. To isolate and identify osteoblast precursor cells using this protocol, *see* **Note 2**.

3.3 Isolation of Osteomacs from Neonates

1. Prepare the collagenase solution and keep it at 37 °C for 30 min before adding to the calvaria at **step 14**.
2. In a biosafety cabinet, prepare a sterile stage by wrapping a sterile board with sterile absorbent towels.
3. Add 30 mL of 70% EtOH to each of three sterile 50 mL beakers.
4. Euthanize 2–3-day-old mouse pups according to institution-approved protocols. One litter of 6–8 pups is adequate for one set of samples including all the controls.
5. Using forceps, dip the pups sequentially in the three beakers of 70% EtOH and decapacitate the pups.
6. Fix the head to the sterile board using T-pins inside the biosafety cabinet.
7. Using forceps, lift the skin from the backside of the head. Insert scissors slowly under the skin and separate the skin from the calvaria by moving the scissors over the calvaria. Cut the skin near the ear on each side and in the middle and remove the skin.
8. Lift the calvaria by forceps and insert scissors under it and cut near the ears on both sides.
9. Lift the separated calvaria slightly and cut near the nose in order to separate it completely.
10. Remove any soft tissue from the calvaria using fine forceps.
11. Cut the calvaria into half and put them in a 50 mL tube containing 10 mL of 4 mM EDTA and incubate for 10 min at 37 °C while shaking in a water bath shaker.
12. Remove the EDTA solution using a Pasteur pipette, add 5 mL of sterile PBS, and shake the tube. Remove PBS using a Pasteur pipette.
13. Repeat **steps 11** and **12** once.

14. Add 5 mL of collagenase solution and incubate for 10 min at 37 °C while shaking in a water bath shaker.
15. Remove the collagenase solution using a Pasteur pipette, discard, and add 5 mL of sterile PBS and shake the tube. Remove PBS using a Pasteur pipette and discard.
16. Repeat **steps 14** and **15** once.
17. Add 5 mL of collagenase solution into the tube and incubate for 15 min at 37 °C while shaking in a water bath shaker. Vortex the tube at an interval of 5 min.
18. In a 50 mL centrifuge tube, add 25 mL complete MEM-alpha. Keep the tube on ice with a 70 μm cell strainer on top of it inside the biosafety cabinet.
19. After 15 min, remove the collagenase and add it to the 50 mL tube through the strainer. After adding the collagenase, add 2 mL of complete MEM-alpha through the strainer to rinse it.
20. Repeat **steps 17** to **19** twice for a total of three digestions.
21. Centrifuge the 50 mL tube at 500 × *g* for 6 min at 4 °C and discard the supernatant.
22. Resuspend the cell pellet in 10 mL of complete MEM-alpha and filter through 70 μm cell strainer into a fresh 50 mL tube. Centrifuge the tube at 500 × *g* for 6 min and discard the supernatant.
23. Resuspend the cell pellet in 500 μL of stain-wash buffer and count using a hemocytometer or automated cell counter.
24. To isolate and identify osteoblast precursor cells using this protocol, *see* **Note 2**.

3.4 Isolation of Bone Marrow Macrophages from Neonates

1. The complete IMDM medium and 1× RBC lysis buffer should be brought to room temperature prior to use.
2. In three sterile 50 mL beakers, add 30 mL of 70% EtOH.
3. Euthanize 2–3-day-old mouse pups according to institution-approved protocols. One litter of 6–8 pups is sufficient for one set of samples including all the controls.
4. Using sterile forceps, dip the pups sequentially in three beakers of 70% EtOH and decapacitate the pups.
5. Cut the hind limbs and keep them in IMDM. If isolating osteomacs and macrophages simultaneously, perform this step after **step 5** from Subheading 3.3.
6. With fine forceps, separate skin and muscles from bones. Hold near the toes with forceps. With another fine forceps, pull the bone out of the skin. Use a sterile sponge gauze to separate muscles from the bone. Cut the bones into small pieces and

keep them in a sterile mortar containing 5 mL of complete IMDM medium.

7. In a sterile mortar, crush the bones gently using a sterile pestle to release the marrow within the bones (*see* **Note 3**).
8. Add 5 mL of complete IMDM medium to the pestle and using a sterile 10 mL pipette, mix thoroughly.
9. Collect the supernatant into a 50 mL tube by passing the cell suspension through a 70 μm cell strainer.
10. Add 5 mL of media to the mortar, rinse the bone fragments, and collect the supernatant.
11. Centrifuge the tube at 500 × *g* for 6 min at 4 °C and discard the supernatant.
12. Incubate with 1 mL of 1× RBC lysis buffer at room temperature for 3–4 min.
13. Add 10 mL of complete IMDM medium to the tube and centrifuge the tube at 500 × *g* for 6 min at 4 °C and discard the supernatant.
14. If the lysis is incomplete, repeat **steps 11–13**.
15. Resuspend the cells in 400 μL of stain-wash buffer.
16. Count the cells using a hemocytometer or an automated cell counter.
17. Resuspend the cells at a final concentration of 2.5×10^7/mL.

3.5 Immunostaining

1. For bone marrow macrophages, 4×10^6 cells for each sample should be stained. 2.5×10^6 cells for each sample should be stained for identification of osteomacs. However, in adult mice, if the yield is low after digestion of bones (due to expression of any transgene or age), a lower number of cells can be used. For single-color controls and FMOs, use $4 \times 10^5 - 1 \times 10^6$ cells (depending on the yield) for each tube.
2. For staining of bone marrow-derived macrophages and osteomacs, the following tubes will be required for each cell types (*see* **Notes 4** and **5**):
   (a) Unstained cells
   (b) CD45 single-color control
   (c) F4/80 single-color control
   (d) CD11b single-color control
   (e) CD68 single-color control
   (f) Ly-6G single-color control
   (g) Mac2 single-color control
   (h) M-CSFR single-color control
   (i) CD166 single-color control

(j) CD45 FMO
(k) F4/80 FMO
(l) CD11b FMO
(m) CD68 FMO
(n) Ly-6G FMO
(o) Mac2 FMO
(p) M-CSFR FMO
(q) CD166 FMO
(r) Sample tube containing CD45, F4/80, CD11b, CD68, Ly-6G, Mac2, M-CSFR, and CD166.

3. Add 1.0 μg of TruStain FcX per million cells and incubate for 10 min at 4 °C.
4. Add the fluorochrome-conjugated antibodies to the appropriately labelled tubes. We recommend titrating fluorochrome-conjugated antibodies before using. For all the experiments described here, 0.5–1.0 μg of conjugated antibodies was used per million cells.
5. As fluorochromes are photosensitive, avoid exposing the cells to light while staining. While adding fluorochrome-conjugated antibodies, turn off the light of the biosafety cabinet. During incubation, wrap the tubes in aluminum foil to avoid exposing them to light.
6. After adding the antibodies, incubate in the dark at 4 °C for 30 min. Vortex the tubes at an interval of 5 min.
7. Add 2 mL of stain-wash buffer, centrifuge the tubes at 500 × *g* at 4 °C for 6 min, and discard the supernatant.
8. Resuspend the cells with 500 μL of stain-wash buffer.
9. Acquire the cells using a flow cytometer (*see* **Note 6**). Alternatively, the cells can be fixed and stored in dark at 4 °C for future acquisition (*see* **Note** 7).

3.6 Acquisition and Analysis

After finishing the above steps, the samples are ready for acquisition. These samples can be acquired using BD LSR II, BD LSRFortessa, or any other instrument with appropriate lasers and detectors. To acquire sufficient events to analyze for bone marrow macrophages, we recommend collecting three million events; for osteomacs, 1–1.5 million events are recommended for each sample. We use the FlowJo software for analyzing the data. At first step, based on forward scatters (FSC-A vs. FSC-W), we use a non-rectangular gate to exclude doublets. Next, on the gated cells, we apply another non-rectangular gate based on forward scatter and side scatter to exclude cell debris. Following that, we sequentially gate the populations in bone marrow macrophages and osteomacs isolated from

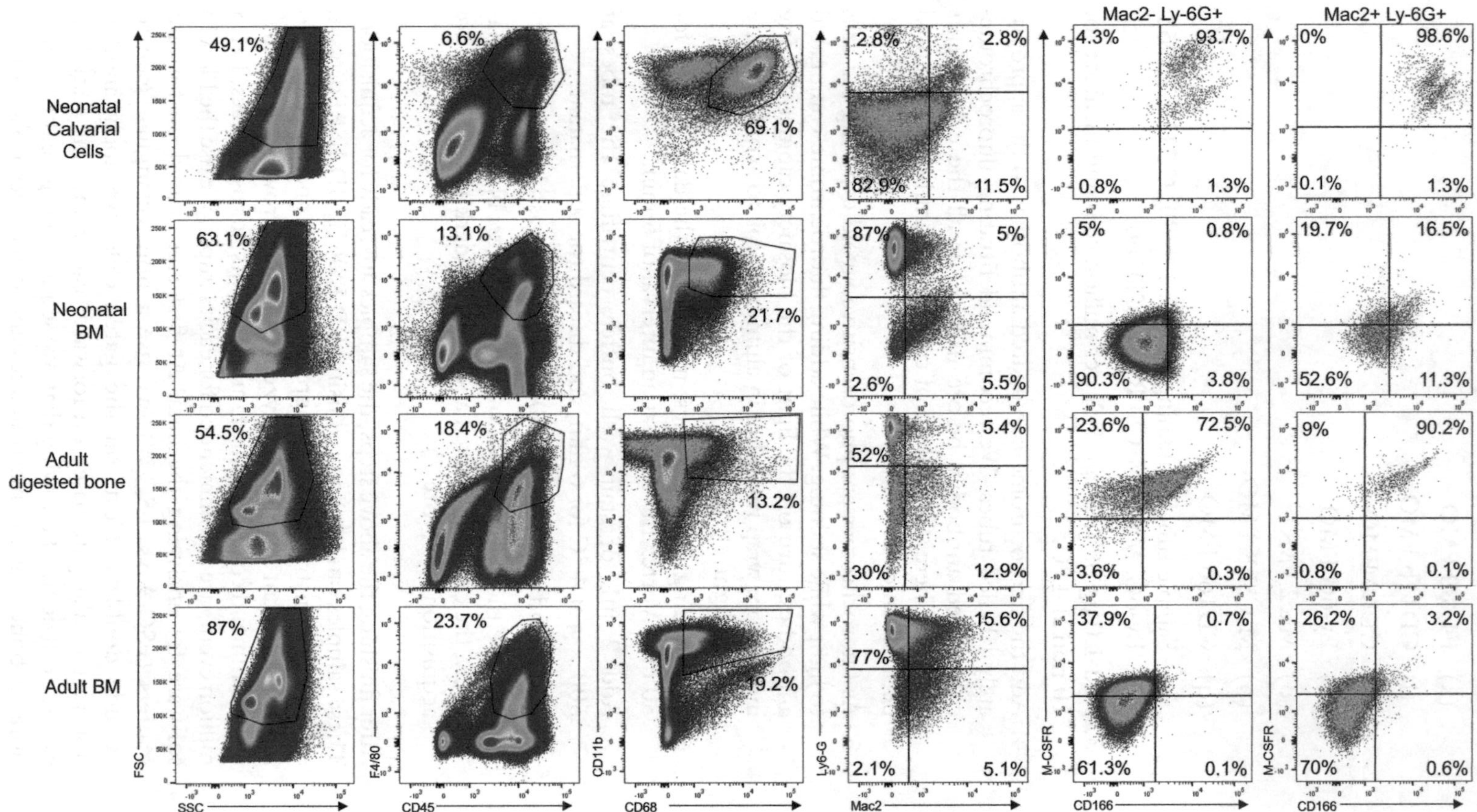

Fig. 1 Representative dot plots of phenotypic analysis of osteomacs and bone marrow-derived macrophages from neonatal and adult mice. Flow cytometric analysis of neonatal calvarial cells, neonatal bone marrow, adult bone, and adult bone marrow. Dot plots are gated according to FMOs and display similarly gated events (left to right). Prior to gating on left-most plot for live cells, we eliminate doublets using a FSC-A vs. FSC-H gate. Phenotypically, we define osteomacs (in neonatal calvarial cells and adult bones) as CD45+ F4/80+ CD11b + CD68+ Ly-6G+ Mac2^{lo} M-CSFR+ CD166+ or CD45+ F4/80+ CD11b + CD68+ Ly-6G+ Mac2− M-CSFR+ CD166+. Bone marrow macrophages (neonatal and adult bone marrow) are defined as CD45+ F4/80+ CD11b + CD68+ Ly-6G+ Mac2+ M-CSFR+ CD166−

neonates and adult mice as shown in Fig. 1. We determine the gates via fluorescence minus one (FMO) controls (*see* **Note 8**). As osteomacs are considered Mac2− or have low expression of Mac2 compared to bone marrow macrophages, we considered both fractions as osteomacs [3, 4, 14, 15]. We defined osteomacs (derived from neonatal calvarial cells or adult bones) as CD45+ F4/80+ CD11b + CD68+ Ly-6G+ Mac2^{lo} M-CSFR+ CD166+ or CD45+ F4/80+ CD11b + CD68+ Ly-6G+ Mac2− M-CSFR+ CD166+, whereas bone marrow macrophages (neonatal or adult bone marrow) are CD45+ F4/80+ CD11b + CD68+ Ly-6G+ Mac2+ M-CSFR+ CD166−.

4 Notes

1. Prepare the collagenase solution fresh on the day of experiment.
2. Subheadings 3.2 and 3.3 can be used to isolate osteoblast precursor cells from adult and neonatal mice. In both instances, CD45− Ter119− CD31− Sca1− cells will be osteoblast precursor cells.
3. Crush the neonatal bones gently to avoid release of osteomacs from long bones.
4. We have validated the 8-color antibody panel described here multiple times in our lab. In principle, the fluorochromes used in this panel can be switched among the antibodies listed here, or an antibody used here can be substituted with a different fluorochrome compatible with the panel and flow cytometer to be used for acquisition. However, we have observed variations in binding capacity of same antibody (and same clone) conjugated with different fluorochromes, even when purchased from same vendor. We have also observed lot-to-lot variation between antibodies conjugated with same fluorochromes. We suggest validating and titrating any new antibody with appropriate positive and negative controls before using for experiments.
5. A viability dye can be included in the panel, preferably conjugated with a Brilliant Violet fluorochrome. If the cells are being fixed for later acquisition, use a fixable viability dye.
6. To prevent the flow cytometers from clogging, we suggest filtering the cell suspensions into a 5 mL tube with 40 μm cell strainer cap prior to acquisition.
7. This protocol uses multiple antibodies with tandem dyes. If the samples need to be fixed for later acquisition, we recommend using fixing buffers appropriate for use with tandem dyes.
8. As the scatter and fluorescent profile varies between cell types, use cells from the same cell types while preparing samples for FMOs. While determining the spread of fluorescence and creating the appropriate gates, we also recommend using FMOs

from same cell types (e.g., for neonatal macrophages, FMOs with neonatal macrophages should be used, not adult macrophages).

Acknowledgments

The authors thank Indiana University Melvin and Bren Simon Cancer Center Flow Cytometry Resource Facility (FCRF) for their outstanding technical help and support. FCRF is partially funded by National Cancer Institute grant P30 CA082709 and National Institute of Diabetes and Digestive and Kidney Diseases grant U54 DK106846. We also thank the support of the NIH instrumentation grant 1S10D012270 for partial funding of the FCRF.

References

1. Ginhoux F, Schultze JL, Murray PJ, Ochando J, Biswas SK (2016) New insights into the multidimensional concept of macrophage ontogeny, activation and function. Nat Immunol 17(1):34–40. https://doi.org/10.1038/ni.3324
2. Ludin A, Itkin T, Gur-Cohen S, Mildner A, Shezen E, Golan K, Kollet O, Kalinkovich A, Porat Z, D'Uva G, Schajnovitz A, Voronov E, Brenner DA, Apte RN, Jung S, Lapidot T (2012) Monocytes-macrophages that express alpha-smooth muscle actin preserve primitive hematopoietic cells in the bone marrow. Nat Immunol 13(11):1072–1082. https://doi.org/10.1038/ni.2408
3. Mohamad SF, Xu L, Ghosh J, Childress PJ, Abeysekera I, Himes ER, Wu H, Alvarez MB, Davis KM, Aguilar-Perez A, Hong JM, Bruzzaniti A, Kacena MA, Srour EF (2017) Osteomacs interact with megakaryocytes and osteoblasts to regulate murine hematopoietic stem cell function. Blood Adv 1 (26):2520–2528. https://doi.org/10.1182/bloodadvances.2017011304
4. Alexander KA, Chang MK, Maylin ER, Kohler T, Muller R, Wu AC, Van Rooijen N, Sweet MJ, Hume DA, Raggatt LJ, Pettit AR (2011) Osteal macrophages promote in vivo intramembranous bone healing in a mouse tibial injury model. J Bone Miner Res 26 (7):1517–1532. https://doi.org/10.1002/jbmr.354
5. Winkler IG, Sims NA, Pettit AR, Barbier V, Nowlan B, Helwani F, Poulton IJ, van Rooijen N, Alexander KA, Raggatt LJ, Levesque JP (2010) Bone marrow macrophages maintain hematopoietic stem cell (HSC) niches and their depletion mobilizes HSCs. Blood 116(23):4815–4828. https://doi.org/10.1182/blood-2009-11-253534
6. Bowen MA, Aruffo A (1999) Adhesion molecules, their receptors, and their regulation: analysis of CD6-activated leukocyte cell adhesion molecule (ALCAM/CD166) interactions. Transplant Proc 31(1–2):795–796
7. Chitteti BR, Bethel M, Kacena MA, Srour EF (2013) CD166 and regulation of hematopoiesis. Curr Opin Hematol 20(4):273–280. https://doi.org/10.1097/MOH.0b013e32836060a9
8. Swart GW (2002) Activated leukocyte cell adhesion molecule (CD166/ALCAM): developmental and mechanistic aspects of cell clustering and cell migration. Eur J Cell Biol 81 (6):313–321. https://doi.org/10.1078/0171-9335-00256
9. Chitteti BR, Cheng YH, Kacena MA, Srour EF (2013) Hierarchical organization of osteoblasts reveals the significant role of CD166 in hematopoietic stem cell maintenance and function. Bone 54(1):58–67. https://doi.org/10.1016/j.bone.2013.01.038
10. Chitteti BR, Kobayashi M, Cheng Y, Zhang H, Poteat BA, Broxmeyer HE, Pelus LM, Hanenberg H, Zollman A, Kamocka MM, Carlesso N, Cardoso AA, Kacena MA, Srour EF (2014) CD166 regulates human and murine hematopoietic stem cells and the hematopoietic niche. Blood 124(4):519–529. https://doi.org/10.1182/blood-2014-03-565721

11. Mair F, Prlic M (2018) OMIP-044: 28-color immunophenotyping of the human dendritic cell compartment. Cytometry A 93 (4):402–405. https://doi.org/10.1002/cyto.a.23331
12. Staser KW, Eades W, Choi J, Karpova D, DiPersio JF (2018) OMIP-042: 21-color flow cytometry to comprehensively immunophenotype major lymphocyte and myeloid subsets in human peripheral blood. Cytometry A 93 (2):186–189. https://doi.org/10.1002/cyto.a.23303
13. Autengruber A, Gereke M, Hansen G, Hennig C, Bruder D (2012) Impact of enzymatic tissue disintegration on the level of surface molecule expression and immune cell function. Eur J Microbiol Immunol (Bp) 2 (2):112–120. https://doi.org/10.1556/EuJMI.2.2012.2.3
14. McCabe A, MacNamara KC (2016) Macrophages: key regulators of steady-state and demand-adapted hematopoiesis. Exp Hematol 44(4):213–222. https://doi.org/10.1016/j.exphem.2016.01.003
15. Wu AC, Raggatt LJ, Alexander KA, Pettit AR (2013) Unraveling macrophage contributions to bone repair. Bonekey Rep 2:373. https://doi.org/10.1038/bonekey.2013.107

Index

A

B

C

Kursad Turksen (ed.), *Stem Cell Niche: Methods and Protocols*, Methods in Molecular Biology, vol. 2002,
https://doi.org/10.1007/978-1-4939-9508-0, © Springer Science+Business Media, LLC, part of Springer Nature 2019

I

M

T

V

X